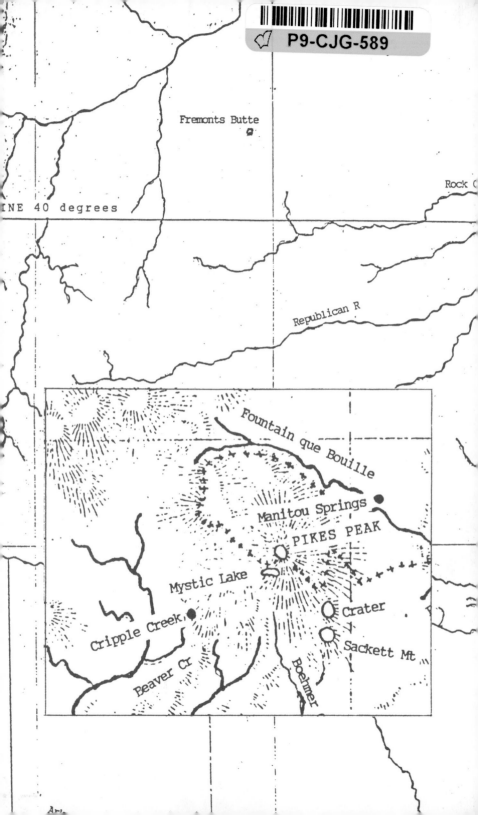

Fremonts Butte

Rock C

INE 40 degrees

Republican R

Fountain que Bouille

Manitou Springs

PIKES PEAK

Mystic Lake

Cripple Creek

Crater

Sackett Mt

Beaver Cr

Boehmer

Ar

Weather Pioneers

ARTIST JOHN A. RANDOLPH'S sketch of the new Signal Service station for *Harper's Weekly*, November 8, 1873. The telegraph line comes up from the south, the flag greets the morning sun, the wind vane revolves at the north end of the building, and the Continental Divide looms behind the peak.

Weather Pioneers

THE SIGNAL CORPS STATION AT PIKES PEAK

By Phyllis Smith

Phyllis Smith

SWALLOW PRESS

OHIO UNIVERSITY PRESS

ATHENS

© Copyright 1993 by Phyllis Smith
Printed in the United States of America

All rights reserved

Swallow Press/Ohio University Press books are printed on acid-free paper∞

Library of Congress Cataloging-in-Publication Data

Smith, Phyllis.
    Weather pioneers : the Signal Corps station at Pikes Peak / by
Phyllis Smith.
        p.   cm.
    Includes bibliographical references and index.
    ISBN 0-8040-0969-4. — ISBN 0-8040-0970-8 (pbk.)
    1. Meteorological stations—Colorado—Pikes Peak—History.
I. Title.
QC875.U72P557   1993
551.65788′56—dc20   92-39291
                                                                          CIP

Contents

ACKNOWLEDGMENTS

THE GREATEST CONTRIBUTORS to the writing of this book are the noncommissioned officers of the U.S. Signal Service, whose logs and journals of the 1870s and 1880s, written while on duty at the summit of Pikes Peak, provide fascinating details on the life of a typical signalman stationed at the top of a very high mountain. Most of them were in their early twenties and relatively inexperienced, but they wrote with style and good humor. Some of their descriptions of rainbows and comets, halos and coronas, and clouds and storms are close to poetry. They remain fresh after more than one hundred years.

Kathryn Shapley of Boulder, Colorado, brought this material to my attention. During the 1960s, she tried unsuccessfully to save the old weather station. She and I agree that the story of these weather pioneers has not been available in book form. In addition to Mrs. Shapley, I am grateful to others who read and critiqued the manuscript: the late Walter Orr Roberts, former director of the National Center for Atmospheric Research; historian Maxine Benson; Michael Becker, Professor of English, Montana State University at Bozeman; and Mary Jane DiSanti, owner of the Country Bookshelf, also in Bozeman.

The historical writer's mainstay is the archivist and the reference librarian. Marjorie H. Ciarlante, Civil Archives Branch, the National Archives, has been extremely helpful, as was James A. Steed, Smithsonian Institution. Archivists Lois Anderton, Carnegie Branch Library for Local Research, Boulder, Colorado, and Sharron G. Uhler, Starsmore Center for Historical Research, Colorado Springs Pioneers Museum, assisted me, as did librarians Mary M. Davis, Pikes Peak Library District, Colorado Springs; Barbara Neilon, Charles Leaming Tutt Library, Colorado College, Colorado Springs; Philip Panum, Department of Western History, Denver Public Library; and Vicky York, Government Documents, Montana State University at Bozeman. Professor Joseph Caprio, also of Montana State University, directed me to technical materials. The thoughtful comments of Bozeman resident and military historian Chris Manos on the modern Army Signal Corps were helpful.

1.

A climb to the summit. High altitude stations.
A weather service is needed. The Signal Service is established.
A station for Pikes Peak. Tall tales.

I N THE LATE SUMMER OF 1873, two U.S. Army Signal Service enlisted men laboriously climbed to the top of Pikes Peak, Colorado. They passed boiling springs and deep mountain lakes; they crossed small creeks, beds of clattering shale, and drifts of last winter's snow. They shivered through rain, hail, and fierce Rocky Mountain winds. The men had been ordered by the chief signal officer back in the States* to determine if a high-altitude weather observatory could be established at that stormy location, more than 14,000 feet above sea level. The signalmen reported to the headquarters of the service in Washington that, despite the wild extremes of climate at the summit, such an installation was possible.

Therefore, for the next fifteen years or so, the Signal Service did operate a tiny year-round weather station on Pikes Peak, the highest observation post in the world at that time and the second mountain facility in the nation, the first having been established in 1870 at the summit of New Hampshire's Mount Washington, 6,288 feet above sea level. Mount Washington was selected because early meteorologists suspected that the mountain lay in the path of most major North American storm patterns, a site for fruitful weather research. It turned out that they were right. And since Pikes Peak lay far to the west of the major populated areas of the United States, its summit was a good location from which to spot eastward-moving storms. Furthermore, not only could researchers observe large land masses in all directions at that height, but also

---

*Colorado was not admitted to the union until 1876.

they felt it was common sense to get up high if they wanted to learn about how the velocity and direction of winds might affect weather conditions and how temperature changes in the atmosphere might exhibit an annual pattern that would aid in weather prediction.

At that time, Austria maintained a weather observation post on Mount Sonnblick at 10,154 feet. Researchers could study weather patterns in India as well, high in the Kashmir in the Ladakh district at 11,503 feet. But nowhere in the world was there a repository of precise weather information upon which to forecast the arrival of wind, rain, and hail storms, or the appearance of hard frost or intense heat. The growing number of meteorologists in the United States welcomed the two high-altitude research opportunities presented by weather stations on Mt. Washington and Pikes Peak, even though they were under military control.

The U.S. Congress was being pressured across the country by two large groups of citizens who wanted a weather bureau set up. Those in the shipping industry along the Atlantic and Gulf coasts and around the Great Lakes were losing men and goods from unexpected and severe storms. They felt their losses might be mitigated if they knew of approaching bad weather ahead of time. Farmers across the nation desired precise weather prediction as well so that they might plant and harvest at more favorable times. The few privately kept weather logs and journals of the 1600s and 1700s had grown in number and some sophistication by the 1860s. However, no amount of weather data could be useful if there were too few experts to interpret the material to forecast changes in the weather. Nor could those predictions be available to citizens across the country if there were no efficient method to spread the word.

Established in 1860, the U.S. Army Signal Service had laid telegraph wire throughout much of the eastern United States by the time of the Civil War. After hostilities ceased, Congress continued to charge the Signal Service with the responsibility of extending telegraph lines toward the West. Added to that charge was the order to establish weather stations around the nation. Thus, weather information would soon be transmitted by wire simultaneously throughout the country and the Signal Service would train observers to man the growing number of weather stations.

The Signal Service has had a long history of developing military communication systems with less than favorable funding and with considerable danger to its personnel. The signaling of enemy troop move-

A SIGNALMAN at work. *(National Archives.)*

ments by means of flags and other devices was the first concern of the Signal Service. Because the signalers had to climb to high places to send messages or to observe the enemy, they were often in danger of falling or being shot at. The improvement of military balloons, dirigibles, and, eventually, airplanes, can be credited to this branch, not without some danger, of course, to the men. The service was on hand in Cuba during the Spanish-American War, and at least one manned observation balloon was shot down by the enemy. The use of carrier pigeons, further refinements to telegraph and telephone, and, later, radio were the special province of the Signal Service. The camera, no longer a crude instrument at the time of the Civil War, occupied the attention of the military as well. In one way or another, the increasing use of all these devices led to the opening of the West.

Despite the apparent lack of public attention, the Signal Corps (as the service was later called) continued to research and refine communication systems. In the early 1970s, the corps moved from Fort Monmouth, New Jersey, to Fort Gordon, Georgia. The Signal Corps of the twentieth century had become a sophisticated arm of the military by the time of the American involvement in the Persian Gulf War. Although the corps has had a distinguished record for the past hundred and thirty years, its accomplishments as a pioneer in communication fields have been taken for granted, says military historian Chris Manos.

The noncommissioned officers who staffed the lonely, wind-blown research station at Pikes Peak were minimally trained for the job at Fort Whipple (later known as Fort Myer), Virginia. True, they were taught to read a thermometer and how to interpret a rain gauge, a barometer, a hygrometer, and an anemometer; however, they were pioneers in the sense that no one in those days had collected enough data to be certain about winds and their behavior, which cloud formations produced rain, hail, or snow, and how lightning and other electrical phenomena might change the weather.

The reports from Pikes Peak in 1873, although sketchy and tentative at first, were transmitted by the signalmen along a newly installed mountain telegraph line and were the object of great interest, even intense fascination, to those who lived below in Colorado Springs, as well as to residents back in the States, where feature stories about the new weather station filled the newspapers. The men reported high and low temperatures for the day and how much rain or snow had fallen; these figures duly appeared in the *Monthly Weather Review*, published by the Signal Service. However, such weather information coming from a 14,110-foot mountain peak, no matter how curious and fascinating, could hardly be useful for daily weather forecasting. Data collected on other subjects, however, were of great use to scientists in the Washington office and elsewhere. The researchers, as well as the public, learned of fierce hail, snow, and thunderstorms in addition to the sightings of halos, coronas, and comets. The observers also told the public of zodiacal lights, Bishop's rings, St. Elmo's fire, and a variety of other electrical mysteries that they experienced on the summit. With each report from the peak, as well as those from Mount Washington, meteorologists everywhere learned more and more about the peculiarities of the winds and temperature changes in the atmosphere.

Perhaps the unusual attention paid these first bulletins inspired some of the Signal Service enlisted men to relieve the tedium of living alone at the top of a mountain by elaborating here and there about the reports. When the public responded enthusiastically to each bulletin from the peak, some of the more imaginative men devoted themselves to the spirited production of outrageous tall tales, an already popular literary form of the 1870s. The Pikes Peak tales were seldom disavowed by residents of the area; they joined together in a silent compact to support and "believe" the romance that "rained down from the weather station," as historian David Lavender put it.[1] One of the Signal Service observers, John Timothy O'Keeffe,* had a particularly fertile imagination. When the weather logs were done, he grew restive with everyday chores and wrote of his venerable mule Balaam, of the heat from a nearby volcano that melted winter snow and ice, and of his encounter with a herd of seven hundred deer and six mountain lions. Sergeant O'Keeffe's greatest story-telling triumph, however, was a collaboration with one Eliphalet Price, a sometime territorial justice of the peace. During O'Keeffe's off hours, as the two men refreshed themselves in a number of Colorado City saloons, they concocted an account of the signalman's "tragic" fight with a plague of pack rats that swarmed across the highest reaches of Pikes Peak.

Lavender wonders if these imaginative tales of Pikes Peak adventures might have encouraged some western migration:

> How many people were moved by these various outpourings to seek new lives in the West is impossible to gauge. It may have been an appreciable total, for the vibrant descriptions, however fanciful, served to confirm what many a restless individual was already feeling. Release, opportunity, and excitement lay out yonder. As the belief mounted, the boom years of the Rockies began to roll.[2]

It is not known whether or not any of the more fanciful reports from Pikes Peak reached members of the U.S. Congress, but in 1891, after years of bickering over whether or not the weather bureau should be under military or civilian control, Congress transferred the responsibility for weather prediction from the Army Signal Service to the civilian Department of Agriculture. The change was due, in part, to the bad behavior of a few high-ranking Signal Service officers in Washington

---

*O'Keeffe's name was often spelled O'Keefe, but his widowed mother signed his induction papers on April 29, 1874, with the name Catharine O'Keeffe. O'Keeffe himself spelled it either way to suit himself.

rather than to deficient weather data from the field. In fact, the success of such station reports was largely responsible for the transition of the weather bureau from a relatively simple military operation to a sophisticated country-wide civilian service. According to Joseph Hawes, "Because of its progressive attitude toward science, the Army weather service sowed the seeds of its own destruction and brought an end to the military control of the weather bureau."[3] For whatever reason, justified or not, just fifteen years after the tiny mountain weather station was established, it was allowed to become obsolete and eventually was closed. The building itself was torn down in the early 1960s and replaced by a tourist reception center. The history of the Pikes Peak weather post, then, is short, but filled with danger, wonder, and excitement, both real and fabricated.

The weather men on Pikes Peak kept careful logs of their observations on the behavior of storms and the patterns of winds at high altitudes—material later published by Harvard University.[4] Although they lacked sophisticated technology and the precision of specialized instruments, these early weather pioneers learned enough to make tentative generalizations concerning the dynamics of the earth's atmosphere, thus surely contributing to the bases for contemporary meteorology. Today, a number of high-altitude research organizations dot Colorado's Front Range. Not only should modern scientists give homage to early record keepers and meteorologists, but they might also pay their respects to a group of young enlisted men of the U.S. Signal Service, short on advanced education but long on personal courage and fortitude.

2.

Some Pikes Peak geography. High-country winds.
The Utes and their forts. Spanish exploration.
Zebulon Pike climbs a mountain. Stephen Long and Edwin James.
What Frémont saw. Ruxton and the springs.
A bloomer girl on Pikes Peak. A camp at Colorado City.

PIKES PEAK RISES ABRUPTLY from the western edge of Colorado's high flat plains, standing watch at the edge of the Rocky Mountains.* It dominates the nearby foothills and seems even more massive than its 14,110 feet because there are no other equally high peaks for fifty miles in any direction, although thirty-one mountains in Colorado have a greater elevation. Officially, Pikes Peak covers 450 square miles. Its slopes are steep; few level spots afford rest for the hiker, who must clamber over loose pieces of granite to get to the summit. Dr. Edwin S. James, botanist for the expedition of Major Stephen H. Long in July 1820, described his first view of Pikes Peak, "the lower half thinly clad with pines, junipers, and other evergreen trees; the upper half a naked conic pile of yellowish rocks, surmounted here and there with broad patches of snow."[1]

On a clear day, when Pikes Peak sparkles in the sun, the visitor at the top can look down to see Colorado Springs at the base of the mountain, and look east to view the great plains stretching into Kansas. Denver is some seventy miles away to the northeast. Beyond Denver, more than 120 miles from the mountain, Greeley is a white speck on the flat prairie, and, beyond Greeley, another white speck is Cheyenne, Wyoming. To the north lies the Front Range of the Rocky Mountains; the snow on Longs Peak glistens in the sun a hundred miles away. To the

---

*The U.S. Board of Geographic Names dropped the apostrophe in Pikes Peak and Longs Peak in 1890.

AN EARLY DRAWING of the east face of Pikes Peak; from J. C. Frémont's *Report of the Exploring Expedition to the Rocky Mountains, 1842.*

west lie the South Park, Middle Park, and North Park basins, and beyond them loom the Sawatch Mountains. Even higher rises the spine of the Continental Divide, more than a hundred miles away. To the southwest, the Arkansas River rushes through the Royal Gorge; the Wet Mountains lie beyond, and even further southwest is the Sangre de Cristo range. Directly south is Raton Pass and the road to New Mexico.

The high winds along Colorado's Front Range have achieved a certain fame, outdone only by Fujiyama in Japan, Mount Washington in New Hampshire, and in areas of the Antarctic. The greatest velocities have been recorded during the winter months (between October and March, with January being the windiest); however, Front Range winds may gust to over a hundred miles per hour during nine months of the year.

Long before Signal Service representatives journeyed out from the States to determine whether or not an observatory could be established at the top of Pikes Peak, others had seen the great mountain, even climbed it. Surely, prehistoric hunting parties circled the mountain to look for deer, antelope, and bear. Later, as evidenced by the arrowheads they left behind, Ute mountain tribes made the area their home for several centuries. Whether the Utes spent much time at the top of the Shining Mountain, as they called it, is hard to determine, but they did establish little stone ledges below the peak to watch for game or to spy on their historic enemies, Comanche or Kiowa hunters, and later, bands of Cheyenne and Arapaho. They also wanted to protect their special high country ground in South Park, earlier called Bayou Salade, from encroachment. These little forts, which could accommodate two or three Ute observers, stretched north of the mountain from Fountain Creek to Bear Creek, near the ascent to Ute Pass.

Francisco Vásquez de Coronado was one of a number of Spanish explorers who wandered as far north as the area of Pikes Peak, which he called Santo Domingo. Soldiers from the Spanish settlements to the south came north again and again to Santo Domingo, looking for their Indian slaves who had run away. Explorer and Spanish official Juan Bautista de Anza was tired of chasing runaway Indian slaves on the plains and led a party farther north to the south face of the mountain in order to circle around and catch what he regarded as uncooperative chattels. One folktale describing these adventures tells of a Spanish climb to the top of the mountain, where an altar of rough granite was built. No one has ever found the remains of such a religious offertory, however.

Surely, a sprinkling of American mountain men, fur traders, and lone adventurers wandered along the uncharted eastern face of the Snowy Range, as the Rocky Mountains were then known. However, there is little historical record to go on because, as historian Robert Athearn says, these men "were an uncommunicative lot and had no precise information about what they had seen."[2] Some of these early adventurers did refer to the massive peak, however, as Great Mountain or Grand Mountain, hardly imaginative titles.

Early historian W. B. Vickers held a poor opinion of one of the first American soldiers to view the tall mountain. Zebulon Montgomery Pike was, Vickers said, "if not an ignorant, at least a superficial observer."[3] When Lieutenant Pike and his party of twenty men ambled north through Louisiana Territory, the United States' most recent acquisition, purchased from France in 1802, they came to what is now the Arkansas River in the late fall of 1806. On the afternoon of November 15, Pike thought he saw a "little blue cloud" far to the west and halted the men so he could look at the "cloud" through his army telescope. This was Pike's first view of the towering mountain that would later bear his name; however, he called this amazing sight merely Highest Peak.

Pike noted in his journal that the large mountain was probably 18,581 feet high, a 4,471-foot miscalculation. He was determined to climb this

LIEUTENANT ZEBULON M. PIKE saw the mountain but did not climb it. *(Library of Congress.)*

"new" peak and set off with a few hardy soldiers. All were dressed in light summer clothing, highly unsuitable for an approaching severe winter storm. The men struggled waist deep through heavy snow fields to a mountain top only to discover that Highest Peak was still ahead, out of reach. The neighboring "hill" they had ascended is now called Cheyenne Mountain, located southeast of Pikes Peak. Frustrated, Lieutenant Pike

EXPLORER STEPHEN HARRISON LONG ambled near Pikes Peak in 1820. *(Painting by Titian Ramsey Peale; Library of Congress.)*

BOTANIST EDWIN JAMES was the first to set down his impressions of a climb to the summit of Pikes Peak. *(Hayward Cirker, ed. Dictionary of American Portraits. New York: Dover, 1967.)*

gave up mountain climbing at that time and continued on his way to Santa Fe, saying later that "no human being could have ascended its pinical."[4]

Fourteen years later, explorer Major Stephen H. Long camped temporarily at the base of the large mountain so that the party's botanist, Edwin S. James, and two others could try for the peak. Dr. James had read Pike's journals, published in 1810, and was determined to make the climb. When he looked up at the mountain, however, he despaired. "The summit appeared so distant, and the ascent so steep."[5] Dismissing Pike's inflated estimate of 18,000 feet as the peak's elevation, James undercalculated the summit height at no more than 11,500 feet.

Luckily, the weather was mild, and the three climbers went up fast, reaching the summit handily at 4 P.M. on July 14, 1820. In his notes, Dr. James comments that the area was incredibly beautiful; here and there, growing from cracks in the rocks, were tiny, brilliant blue alpine flowers. He described an immense ravine just down from the mountain summit on the southwest side. The men found it curious that so many grasshoppers were flying about the mountain and wondered if some sort of plague was in progress.[6]

EXPLORER JOHN C. FRÉMONT had trouble viewing the peak in 1842 because of forest fires. (*Western Historical Collections, University of Colorado at Boulder.*)

Major Long named the granite mountain James Peak to honor the hardy botanist for his historic climb. For a number of years, either *Pike* or *James* appeared on maps of the area. By the time of Colorado's gold rush in 1858, however, Dr. James had been forgotten and the mountain was called Pikes Peak exclusively.

Before and after the Long expedition, French trappers were often in the area. One of them named the creek that flows to the north of the mountain Fontaine qui Bouille, or the "water that boils," even though it boiled lukewarm.

John Charles Frémont and his party tried to view the mighty peak in 1842, but smoke from a nearby forest fire obscured it. As they proceeded south, they saw the summit clearly for only a moment before clouds, mist, and rain curtained it again. Frémont characterized the area as "an immense and comparatively smooth and grassy prairie in very strong contrast to the black masses of timber, and the glittering snow above them."[7]

Although the Indians were said to worship the springs along Fountain Creek—they called them after their god Manitou—another visitor to the area, travel writer George F. Ruxton, "dismissed" the Indian view. According to early French-Canadian trappers, he said, the springs "are the spots where his satanic majesty comes up from his kitchen to breathe the sweet fresh air, which must doubtless be refreshing to his worship after a few hours spent in superintending the culinary process going on below."[8] Ruxton pondered the possibility of attempting to climb the mountain, considered the energy he would be required to expend, and decided against it. He mentions that, again, the Pikes Peak region was thick with smoke from forest fires in 1847.

Perhaps Zebulon Pike would have been chagrined to learn that, in the late summer of 1858, a slight, twenty-year-old woman reached the top of the mountain that he had predicted no one would ever climb. Julia Archibald Holmes decided to make the ascent well before she and her husband, James, started the trip west from Lawrence, Kansas, with a gold-seeking party. As the ox-drawn wagons rolled across Kansas Territory, Mrs. Holmes strode beside the caravan each day for five weeks. When she was able to walk briskly for ten miles at a time, she felt she was ready for the rugged climb to the summit.

Julia Holmes's hiking costume was a difficult sight for some, a scandal to others. She wore a short calico dress over bloomer pants—an outfit known in some eastern feminist circles as the American Costume. Mrs. Holmes was no doubt correct in saying, "I am perhaps the first woman who has worn the American Costume across the prairie sea which divides the great frontier of the states from the Rocky Mountains." Another woman on the trip, who knew her place and stayed inside a stifling wagon each day, advised Mrs. Holmes: "If you have a long dress with you, do put it on for the rest of the trip, the men talk so much about you." Curious, Mrs. Holmes asked, "What do they say?" The well-meaning woman could not give her a good reply, except to say, "Oh nothing, only you look so queer with that dress on."[9]

The Kansas party reached the base of the great mountain on August 1. Julia and her husband James rested a bit, then started the ascent. On August 5 they passed by what they called a snowy "yawning abyss," the "ravine" that had been described earlier by Dr. James. Julia noted in her journal the little, bright blue flowers growing from green tufts in the granite. As they approached "within a few hundred yards of the

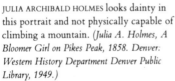

JULIA ARCHIBALD HOLMES looks dainty in this portrait and not physically capable of climbing a mountain. (*Julia A. Holmes, A Bloomer Girl on Pikes Peak, 1858. Denver: Western History Department Denver Public Library, 1949.*)

JULIA ARCHIBALD HOLMES hiked to the peak in "The American Costume." (*A Bloomer Girl on Pikes Peak, 1858. Denver: Western History Department Denver Public Library, 1949.*)

top . . . the surface changed into a huge pile of loose angular stones, so steep we found much difficulty in clambering up them."[10] When the couple reached the final boulder field at the summit, they did not tarry for long. Julia stood in the wind, the American Costume flapping, made a few notes in her journal, and read aloud a short passage from Emerson. (Perhaps it was a favorite verse; it did not seem to have much relation to the grandeur of the scene: "A ruddy drop of manly blood/The surging sea outweighs,/The world uncertain comes and goes;/The lover rooted stays.")[11] As they gazed about, they could see a fierce storm forming, so the couple moved briskly down the mountain to avoid being caught in the snow that quickly blanketed the area.

One year after Julia and James Holmes climbed Pikes Peak, a little brawling camp named El Paso was established along the eastern base of the mountain by a few gold seekers who had come west with the Kansas party and who had decided to stay on. They reasoned that a town built along an Indian trail just east of Ute Pass could not help but flourish. "Despite beautifully lithographed and widely distributed maps which 'emblazoned to the world that a new town had enlarged the area of civilization,' the venture failed."[12] In January 1859 those who were left changed the name of the settlement to El Dorado. By spring, the camp was deserted. Those with the money had departed. What was left was a number of dilapidated cabins built along Fountain Creek.

In 1860, in further hopes of economic boom, the residents rechristened the camp Colorado City. Traveler Henry Villard came through that year, looked over the town of Manitou Springs as well, and sampled the water from Fontaine-qui-Bouille. It tasted good, he said, and reported to his readers that the "inhabitants largely use the water of the soda springs for making bread, etc."[13]

For a time, the territorial legislature seriously considered Colorado City as the site for a permanent capital. However, after four days at the site in 1862, the lawmakers moved back to Denver City when they determined that Colorado City's accommodations and "diversions" were not only inadequate but unseemly. Apparently, some of the camp's more permanent eighty-one residents spent a great deal of time in the many saloons, participating in brawls and shootings.

Gold fever had abated for a time. Back in the States, the Civil War required the capital that might have been used to build railroads and telegraph lines to the Pikes Peak area. In 1864 a great flood roared down Fountain Creek, sweeping away a number of the cabins. One month later, a plague of grasshoppers hit the area. Colorado City was destined to become a ghost town before it got a proper start.

3.

H UMAN SOCIETIES HAVE ALWAYS been preoccupied with changes in the weather. Some have ascribed these changes to the whims of supernatural powers; other societies developed crude tools to observe and measure rainfall, snow level, and the direction and velocity of the winds. By the time of Colonial settlement in the United States, weather watchers generally held that changes in temperature as well as clouds of rain or snow moved across the American continent in a west-to-east direction. Attempts to predict those changes, however, were relatively primitive and scattered until the early 1800s. True, scientists did have weather vanes, nonrecording rain gauges, minimum and maximum thermometers, and mercury barometers. Curious men like Benjamin Franklin worked with those instruments, in addition to his balloons and kites. He concluded from his experiments that weather moved, not in a straight west-to-east pattern, but, more likely, in a southwest-to-northeast route, an important first principle in storm prediction. But more sophisticated observations and predictions had to await the development of more precise instruments and the means to transmit weather information quickly.

Thomas Jefferson maintained such careful weather records at his Virginia plantation that many Americans wrote to him for data from his observations. As president, he required that explorers Meriwether Lewis and William Clark keep weather logs on their expedition to the northwest in 1804-1806. In 1812 Surgeon General of the Army James Tilton directed that all military hospitals under his command keep climatological records. This system eventually grew to a network of ninety-seven

BENJAMIN FRANKLIN'S inquiring mind led him to the observation that storms in the United States generally moved from southwest to northeast. *(James T. White and Company.)*

army bases directed by James P. Espy, who was appointed by the Congress in 1842 as Meteorologist to the United States Government. In 1817 Josiah Meigs, commissioner-general of the U.S. Land Office initiated a simple program whereby its divisions recorded daily precipitation and temperatures. Universities and colleges in the Northeast followed with similar programs. Thus, both military and civilian institutions were gathering weather information but none had a method to quickly send this information to a central location for review.

In 1841 The U.S. Patent Office supervised a small group of weather volunteers who logged their findings and sent them on to Washington by mail. Joseph Henry, secretary of the Smithsonian Institution, was so enthusiastic about this program that he took it over and enlarged it to include one hundred and fifty reporting stations by 1849. For the first

THOMAS JEFFERSON kept meticulous
weather records at Monticello. *(Library of
Congress.)*

JAMES POLLARD ESPY became the nation's
first official meteorologist in 1842.
(Smithsonian Institution.)

time in the United States, Professor Henry used the telegraph to collect
weather data from these posts. He secured the cooperation of the newly
established Western Union; each day, before any commercial messages
were sent, every telegraph office was required to open its lines to the
Smithsonian to answer the question, "Good morning, what is the
weather?"[1] As telegraph lines expanded, so did the Smithsonian network; by 1860, some five hundred observers reported regularly to Professor Henry's group in Washington. The secretary dreamed of the day
when a national weather bureau would operate under the Smithsonian
umbrella, but he was forced to discontinue the project for a time when
hostilities broke out between the North and the South. After the war,
Joseph Henry resumed his reporting network but could not interest
Congress in expanding the Smithsonian operation to include a forecasting service. Nevertheless, he and his fellow scientists continued to push
for such a program.

In France weather forecasting had become more than a notion by
1855. Dutch scientists were participating in a forecasting program by

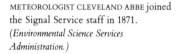

METEOROLOGIST CLEVELAND ABBE joined
the Signal Service staff in 1871.
(*Environmental Science Services
Administration.*)

THE FIRST CHIEF SIGNAL OFFICER of the
Signal Service, Albert J. Myer was
charged with formation of a weather
bureau in 1870. (*U.S. Army.*)

1860; Great Britain followed one year later with a similar service de-
rived from a thirty-station network. With greater use of the electric
telegraph during the Civil War, the daily, even hourly, transmittal of
weather information to the rest of the country seemed a real possibility.
But recording instruments were expensive and observers needed special
training to use them.

Cleveland Abbe was a meteorologist with better training than most.
He had studied with German astronomer F. F. E. Bruenow at the University of Michigan. He worked with B. A. Gould of the Coastal Survey at Cambridge, Massachusetts, in 1859 and 1860 and later spent two
years in Pulkova, Russia, learning from astronomer Otto Struve. In 1868
Abbe became director of the Mitchell Astronomical Observatory in
Cincinnati and started to develop a network of weather observer stations in the Great Lakes area. By the end of 1870, Abbe heard regularly
from thirty-three watchers in Wisconsin and Michigan. When he showed
a colleague his charts of weather data collected by daily telegraph, Pro-

METEOROLOGIST INCREASE LAPHAM saw that a national weather bureau proposal was introduced to the U.S. Congress in 1870. (*Environmental Science Services Administration.*)

fessor Abbe commented, "I have started that which the country will not willingly let die."[2]

Fellow meteorologist Increase Allen Lapham, a resident of Milwaukee who had kept local weather observations since 1827 and had tried without success to set up a state weather bureau, suggested to his congressman, General Halbert E. Paine, via a report entitled "Disasters on the Lakes," that it was time to establish some sort of nationwide weather bureau, based on Professor Abbe's Great Lakes model.[3] Neither Abbe nor Lapham had seriously considered that the responsibility for such a weather forecasting service be placed with a branch of the military. Neither did Professor Henry of the Smithsonian Institution when he wrote to Congressman Paine that "there can be no doubt from the present state of meteorological science that a properly devised system of weather telegrams would be of great importance to the welfare of commerce as well as of much interest to the general public."[4] But these men were scientists and Congressmen Paine and his colleagues were politicians. And Paine had heard from a military man as well.

Albert James Myer, a wealthy doctor from Newburgh, New York, had been involved since 1860 with the Signal Service, a group he helped to establish. Earlier, Myer had studied at Buffalo Medical College, where he wrote a paper entitled "A Sign Language for Deaf Mutes." As a lieutenant with the U.S. Army in 1856, he was posted to New Mexico Territory, where he observed firsthand the Comanche Indian method of signaling one another at a distance. From this borrowed concept of communication, Myer devised a signal system of flags and similar devices that was used during the Civil War. Tall and imposing, Myer was considered impetuous by his fellow officers. He loved new ideas and was impatient with those who waited for official approval before acting. Myer was really a colonel, but most deferred to his brevet Civil War rank of brigadier general. Despite his rocky relations with the military during the war, he was named chief signal officer of the Signal Service, a group that was casting about for postwar jobs and recognition. The service was about to become "as extinct as the dodo," despite its popularity with those Americans who loved colorful flag signals and daring balloon ascents. True, the service was given the responsibility to "equip and manage the field electric telegraph for use with active forces" in 1867, but General Myer needed to solidify the Signal Service position.[5]

He therefore wrote to Congressman Paine:

> By a coincidence, I had caused some maps, showing possible coast telegrams and signal stations, to be arranged for the War Department, before the Congressional Papers concerning the proposed government weather service reached me.
>
> I have been much impressed with the importance of the endeavor proposed in the bill by you, as an aid and safeguard to navigation, and as a mode not before availed of in this country of utilizing in the interests of commerce, the posts and force which must be maintained for military purposes in the interior and upon the sea coast.[6]

The bill for a national weather service was passed by joint resolution on February 9, 1870; it provided "for taking meteorological observations at the military stations in the interior of the continent and at other points in the states and Territories of the United States, and for giving notice on the northern lakes and at the seacoast, by magnetic telegraph and marine signals, of the approach and force of storms."[7]

Although Professor Abbe and his colleagues had hoped for a civilian weather bureau staffed by those with scientific backgrounds, they were delighted that, finally, a national facility for the prediction of thunderstorms, hurricanes, cold waves, and other potentially dangerous weather systems was about to be established. Could the military keep reliable records, they wondered. Perhaps they did not fully grasp the character of General Albert J. Myer. Several months before all this was official, the chief signal officer was training weather station personnel at Fort Whipple, Virginia. Lack of knowledge did not deter the general. True to his reputation for dispatch, Myer worked hard to expand and train his organization without himself having much familiarity with the subject. He learned fast and expected those serving under him to learn fast as well. At Fort Whipple the men took courses in military tactics, signaling, telegraphy, telegraph line construction, electricity, meteorology, and practical observation of meteorological events. They were told that when they went out into the field, they were to become ambassadors as well. They were to call upon businessmen, chambers of commerce, newspaper editors, college professors, politicians, and anyone else who could enhance the reputation of the U.S. Signal Service. By November 1, 1869, three months before official congressional approval, quickly trained sergeants and their assistants were dispersed to twenty-four cities in the nation to set up weather stations as far south as New Orleans and as far west as Cheyenne, Wyoming.

For some months, Chief Signal Officer Myer had been courting Increase Lapham and Cleveland Abbe because he knew he needed their expertise. Myer's attentions must have paid off; late in 1870, after some grumbling, Professor Lapham consented to come to Washington to be the general's assistant; however, the gentle Quaker left again almost immediately, due perhaps to ill health or his distaste for serving a military establishment. Scholarly Cleveland Abbe, who had also grumbled heretofore, took Lapham's place as the nation's official weather forecaster; he left Cincinnati for Washington in January 1871 to be assigned to General Myer's personal staff. It must have been a compelling responsibility, for Abbe stayed there, in one capacity or another, for the next forty-five years. A wedding between science and the military seemed to be successful, at least for a time.

As Cleveland Abbe assumed his forecasting duties, he ordered that no local observer be allowed to predict weather conditions for any area.

He wanted no uninformed wildcat forecasting anywhere in the system. Only Washington had that authority, Abbe decreed, and only Washington would receive by telegraph twice daily the necessary reports from all stations, the data from which could then be plotted on synoptic charts—maps displaying weather conditions over a broad area. Professor Abbe vowed that, forty-nine minutes after telegraphic delivery of data from around the country, he would produce a national weather "probability." The meteorologist was leery of the word *forecast* and preferred *probability* until December 1876, when he decided that the word *indication* was best suited for weather prediction. By 1881 he felt more secure, apparently; *indication* became *forecast*.

Following Abbe's orders to the letter, the sergeants' first weather reports on the wire were simple, a laconic "fair" or "cloudy" with no hint of prediction of future conditions. Very soon, however, the newspapers received synoptic maps and charts containing such "sophisticated" data as local temperature, barometric readings, and wind velocity and direction. One major daily newspaper editor, unfamiliar with the new format, printed the map sideways, to the total confusion of his readers.

By Christmas of 1870 the first high-altitude observatory in the United States had been established on the summit of New Hampshire's Mount Washington under the wing of the Signal Service. Scientists around the nation took note that the government was getting into the research business. Those scientists at the New Hampshire site dubbed it "the birthplace of the winds," as they cracked the ice from their beards.[8] To those citizens who did not have to endure living on an icy, windy mountain top, the new weather information was excitingly romantic. They wanted more. Apparently, so did Congress. In 1871 the War Department published a brochure on weather which explained that

> by increasing the mountain stations, and by adding such balloon observations as can be made, and specially by the study of the forms, changes, motions, height, and velocity of the clouds, and of the optical phenomena of the atmosphere . . . meteorologists hope eventually to arrive at a full knowledge of the regions of the air, where the severe storms are propagated.[9]

An ambitious goal.

It was only a matter of time before Signal Service eyes looked west to the Rocky Mountains for a second high-altitude weather station site.

WILLIAM J. PALMER, builder of railroads and founder of Colorado Springs. *(Starsmore Center for Local History, Pioneers Museum, Colorado Springs.)*

Meanwhile, other eyes, promoters' eyes, were looking over the Pikes Peak area rather carefully with less altruistic goals in mind. In 1869 a thirty-five-year-old railroad financier who was completing construction of the Kansas Pacific Railroad to Denver journeyed south to visit a site at the base of the mountain. General William Jackson Palmer, a Civil War hero from Pennsylvania, decided it was a promising location for a special community, the reputation of nearby Colorado City notwithstanding. He liked the area, despite its 6,000-foot elevation and despite the "blackened waste"[10] from prairie fires, which had charred nearby stands of buffalo grass and yucca. Palmer promptly bought 10,000 acres just east of Pikes Peak for one dollar per acre and capitalized the land as the Fountain Colony. The promoter began construction of the first narrow-gauge railroad in the territory, the Denver and Rio Grande, from Denver to the "new city," a distance of seventy-five miles.

By October 1871 Palmer had completed the line and had already met in Denver with members of the fledgling colony to plan the "ideal community," a scheme he hoped would appeal to the wealthy and select, those of "good moral character and strict temperance habits."[11]

Invalids, especially those with consumption, who wanted to come west for Colorado's pure air, were welcome, if they had sufficient money. The general also hoped to attract his future bride, Queen Mellen, to what he predicted would be a special place to live, a hope which, alas, Queen did not share. After they were married, Mrs. Palmer came west from time to time, but never regarded herself as a permanent resident; she preferred the more civilized life in the East and in London.

After considering and rejecting the name Fountain Valley, Palmer called his new community Colorado Springs. No matter that the springs were located at nearby Manitou; he thought Colorado Springs had a nice sound to it. Soon houses and hotels were under construction. Palmer had not yet given up on Queen and started building his own elegant mansion northwest of the fledgling town; he called it Glen Eyrie.

The existing and decrepit Colorado City was not totally ignored by Palmer; the camp would have its uses. He reasoned that mills, smelters, saloons, gambling halls, and other establishments catering to man's baser nature could best be confined to Colorado City. No liquor was to be sold or manufactured in the new community, a law that was more or less in force until 1933. Later, labor leader William D. "Big Bill" Haywood described the Colorado City camp as a "forlorn little industrial town of tents, tin houses, huts, and hovels . . . nothing but waste and slum . . . bordered by some of the grandest scenery of nature."[12]

Nearby, the new Colorado Springs grew quickly. By the end of the first year of settlement, eight hundred residents called "the Springs" home. One hundred and fifty buildings were up, including several sanatoriums for well-to-do tuberculars. Drinking water was sold in town for twenty-five cents a barrel. "Antelope Jim" Hamlin sold meat to the new residents. Because so many of the townspeople had immigrated from the British Isles, the community was nicknamed Little Lunnon. The newcomers brought their ways with them. Thinking of home, these privileged citizens organized fox hunts, polo games, and cricket matches, even though the traditional green grass for these activities was still a dream. It was not long before six Shakespearean companies flourished

EARLY COLORADO SPRINGS. (*Western History Department, Penrose Public Library, Colorado Springs.*)

in town, a cultural development completely foreign to the more earthy interests of nearby Colorado City residents.

British traveler and journalist Isabella Bird rode through the mountain area on horseback on November 1, 1873, and came down to take a look, but not before she changed from riding pants to flowing long skirts and sidesaddle for town. She was not overwhelmingly impressed with the construction in either community:

> After fording a creek several times, I came upon a decayed-looking cluster of houses bearing the arrogant name of Colorado City, and two miles farther on, from the top of one of the Foot Hill ridges, I saw the bleak-looking scattered houses of the ambitious watering place of Colorado Springs, the goal of my journey of 150 miles.

Most unattractive, she wrote, and moreover, there were no trees! But she loved the colored rocks of the gorges, was awed by the "ghastly peaks," intrigued by General Palmer's hideaway at Glen Eyrie, and delighted by the springs at Manitou. Off she went to discover Estes Park and "Mountain Jim" Nugent.[13]

In 1872 another explorer, noted geologist Ferdinand V. Hayden, flushed with the success of his expedition to the Yellowstone area the previous year, was in the Pikes Peak region to improve the existing maps. He and a group of his men climbed to the summit, moved a number of huge boulders about to form a large cairn, then tramped partway down on the southwest slope to map an extinct crater that Edwin James had earlier called a ravine and Julia Holmes had called an abyss. A stream nearby was named Cripple Creek, an area that in another nineteen years would be filled with gold prospectors. Hayden made an understated note in his log that the volcano area probably was a likely place to find gold. The geologist determined that Pikes Peak was 14,134 feet above sea level—in sharp contrast to Zebulon Pike's estimated 18,581 feet. (Modern geographers list 14,110 feet as the official height for Pikes Peak.)

On May 31, 1873 the newly established *Colorado Springs Weekly Gazette* reported that the Denver papers were full of the news of the Hayden expedition. Moreover, Dr. James T. Gardner of that group announced, to the *Denver News* that Chief Signal Officer Myer of the War Department would send a party sometime in the summer to look over the possibility of establishing a "station of observation" at the summit of Pikes Peak. The only condition for such a visit, Gardner said, would be that the town of Colorado Springs agree to put up some money for a

seventeen-mile telegraph line to the mountaintop. The news gratified
Colorado Springs residents Captain W. W. Allen and rancher R. F.
Song, for they had written to Washington officials a year or so earlier
to plead the case of Pikes Peak as the perfect site for the second mountain observatory in the nation. Local businessmen, seeing the advantage
of a weather station atop Pikes Peak as an attractive lure for tourists,
quickly raised two hundred dollars for the telegraph line; they were
quickly followed by General Palmer's Denver and Rio Grande Railroad, which contributed five hundred dollars.

Meanwhile, General Myer was looking over his roster to determine
which of his men would best represent the Signal Service at the top of a
lonely mountain peak that had the reputation of unpredictable and violent storms, unbelievable cold, and, in general, an impossible location
for human habitation. How could anyone live year-round on that peak?

4.

Visitors from the East. James Smith tells a story.
Another trip to the summit. To build a weather station.
A taste of winter.

I N JULY 1873 SERGEANT George Boehmer and Private James H. Smith
of the U.S. Signal Service arrived in Colorado Springs from Washington, rented a small office, and prepared for their first assault on the summit. They soon learned that there were two possible routes to the peak. One was a steep nine-mile climb from the boiling springs at Manitou. The other way was longer, with an eight-mile hike to Bear Creek Canyon, on to Mystic Lake (later called Lake Moraine), and up through the forest to timberline and rocky terrain.

The two men opted for the longer route and loaded a burro with their heavy coats and a few camping supplies. The weather was fine but, almost immediately, they were forced to stop and rest. Thirty-one-year-old Boehmer, a stocky immigrant from Berlin, was the hardier of the two, but both men were unused to the altitude, despite Smith's duty on Mount Washington the previous spring. By noon, they had only gone as far as the mouth of Bear Creek Canyon, having been forced to ford the creek twenty-one times.

It was fully dark when they reached Mystic Lake. After calculating the elevation with an aneroid barometer, they determined that they had another four thousand feet to climb. The men decided to camp there for the night. James Smith was somewhat apprehensive, as he had never before slept outside in what he called "the wilderness . . . away from the rest of the world, the silence was so profound, the starry sky so sparkling, and the lowering mountains which surrounded the lake so massive and gloomy." Smith finally fell asleep, only to be awakened at 2 A.M. by Boehmer, who calculated that if the two were to reach the summit by sunrise, they must lose no time. Smith said later he was "in

29

GEORGE H. BOEHMER was part of the first crew at the station in 1873. *(Photo by Davis; Smithsonian Institution.)*

favor of postponing the trip, but Boehmer, more enthusiastic, dragged me upon my feet, and away we went, stumbling up the steep path in the darkness. It was a toilsome journey."[1]

They did get to the top, however, shivering and shaking in the cold, raw dawn, teetering on loose shale at the peak. They faced east, where a vague light began to illuminate the horizon. Smith continued to describe their adventure:

> Above our heads the sky was clear and starlit; beneath our feet, stretching out for a hundred miles, lay a bank of snowy clouds looking like a frozen ocean. Away off in the east a narrow rim of

gold appeared, and when a handbreadth of the sun arose the ocean
began to move. The billowy clouds gently rose and fell while
athwart them the sun shot long arrows of gold and silver light.

It was a gorgeous and impressive sight, but it did not last more
than five minutes. The mass of clouds dissipated as the sun rose
higher and higher, and in an incredibly short space of time, the sky
was entirely clear.[2]

The trip down was quick and uneventful, although tiring. When the
men reached their Colorado Springs office, their knees had turned to
jelly and they both fell to the floor, exhausted. As they sent off their
first report to Washington, General Myer was already forming another
party to make a second ascent of Pikes Peak.

On September 6 Private Edward W. Boutelle of the Signal Service
and Officer Fitch of the U.S. Coastal Survey arrived in Colorado Springs
from Washington. C. Ford Stevens of the Denver and Rio Grande
Railway came from Denver; Sergeant Boehmer and Private Smith
completed the party of five "mountaineers."

The second trip was not a total success. Before the men got started,
one of the mules, "wanting a little exercise, and not knowing what was
before him, amused himself by shaking loose his pack consisting of tin
kettles, hatchets, telegraph wire, etc., and galloped about town."[3] On
the trail at last, Sergeant Boehmer had trouble reining his horse and
broke his finger. Boutelle's horse threw him off and the private landed
fifteen feet down a gulley full of rocks; his knee was badly injured and
the new barometer he carried was broken. Both men returned to town
in some pain and embarrassment as the other three men went on their
way.

When the small party reached the summit, they agreed that Pikes
Peak was, indeed, the right place for a weather station, even though the
base of the mountain at Colorado Springs was at a higher altitude than
the top of the station on Mount Washington. They could see over the
tops of nearby mountains for a distance of some fifty miles. Despite the
difficulties at the start of the trip, the men felt that access to the top was
relatively easy compared to other high peaks along Colorado's Front
Range, which extends north to Wyoming, where climbing was hazard-
ous and restricted to hardy mountaineers. They also agreed that the
U.S. Army road builders should improve the trail without delay. A station at Pikes Peak, they decided, was the ideal site for research and
study of atmospheric changes and their relationship to accurate weather
forecasting.

The next bulletin from Washington put Boehmer and Smith in very
low spirits. Chief Signal Officer Myer had noted in their report to him
that some sort of building was needed at the summit if weather observations were to be made. Of course, The general ordered the men to
waste no time in building one. Smith remembered:

> Fancy our predicament! Five men, not one of whom knew any-
> thing about building, all from the East, expected to go up into an
> altitude of fourteen thousand feet and build a house! It was a pre-
> posterous idea, but Boehmer, who was the observer in charge of
> the party, prepared to obey orders. We purchased for each man
> one pick, one shovel, one trowel, one axe, one hammer, and then
> laid in a complete camping outfit for the party, including a large
> "A" tent. The supplies and provisions loaded down five pack
> animals, and as each man was mounted we made quite a cavalcade
> as we started out on our foolish venture.[4]

The men managed to haul the tent up to the summit, but that was the
extent of their construction activities there. Two miles down, below
timberline and near Mystic Lake, they did build a small log cabin, however, where they took residence while Sergeant Boehmer fired off a
number of reports to Washington.

There they stayed at what they called Camp Howgate. Shortly thereafter, Lieutenant H. Jackson, the disbursing officer for the Signal Service, "came out from Washington to see what the trouble was. He went
up on the peak, looked at it for about five minutes, and decided that the
project was absurd. If the house was built at all, it must be by skilled
mechanics."[5] Jackson evidently knew how to cut bureaucratic tape; he
ordered supplies, signed a $2,500 contract with a Denver builder, and
saw that the house on top of the peak was completed by civilian construction specialists in less than four weeks, before the heavy snows.

Despite Lieutenant Jackson's decisive actions, the weather station
was not an easy building to construct:

> Rocks were plentiful but every bit of timber and board had to be
> transported by man pack, horse pack or mule pack. This included
> the transportation of telegraph poles, wires, and batteries, tin
> sheets for the roof, as well as everything that would be needed at
> the station, such as food supplies, pots, pans, dishes, the stove, and
> coal for the stove. Three fat steers were driven up as far as timber-
> line and slaughtered, dressed, and the beef packed by mule to the
> signal house, and stored for winter use.[6]

EARLY VIEW OF THE north wall of the station at Pikes Peak. Note telegraph poles in place and newly installed louvered shelter for hygrometer and temperature equipment to the left of ladder. (*Western History Department, Penrose Public Library, Colorado Springs.*)

Judging by the description of stone mason John Gibson, one of the thirteen men hired to erect the station, it is surprising the building lasted a season:

> As near as I can remember the size of the building was eighteen by thirty feet and about ten feet high. Shed roof, stone laid up dry and rough-dashed afterwards. . . . Most of the trail was covered with decomposed granite and it wrought havoc with our footwear, so much so that we had to wrap our feet in gunny sacks before the job was near completed.[7]

From the beginning, the weather station was inadequate. Even though the stone masonry walls were eighteen inches thick, snow filtered through the cracks and through the poorly fitted windows and doors to pile up inside the two small rooms. One served as an office and the bedroom for the sergeant-in-charge; the other housed the kitchen, storage of equipment, and sleeping quarters for the assistants.

The weather that particular September did not always cooperate with those who were responsible for the improvement of the trail and construction of the signal station. On September 13 James Smith and recently assigned Private Leander A. Lemman got an early taste of what a

Pikes Peak winter was going to be like. On their way from Colorado Springs to the log cabin near the timberline lake, the men were caught in a snowstorm fierce enough to make further climbing impossible. They spent the night huddled under a rock outcropping without benefit of fire, food, or blankets. In the morning, bright sunshine warmed them and they got back down to the Colorado Springs office safely. When they arrived, their colleagues told them that the forthcoming dedication of the signal station on top of the mountain, planned for mid-October, was receiving considerable attention in newspapers throughout the nation.

5.

THE WEATHER OBSERVATION POST was formally dedicated on October 11, 1873. The day was a festive one for the large number of ladies and gentlemen who made the trek from the Springs to the summit on a trail that had been improved in recent weeks by the U.S. Army.

Colorado Springs resident and literary light Grace Greenwood made a pretty little speech, as did other notables. Sergeant Boehmer responded gracefully to their comments. A young lady, also from the Springs, presented a giant American flag to the signalmen. *Harper's Weekly* described the event to its readers: "When the Stars and Stripes were flung to the breeze from the highest signal station in the world, three rousing cheers were given."[1] Many in the party stayed overnight in order to see the sunrise from the top of the mountain. This event was the first of many tourist expeditions to Pikes Peak.

Harper's Weekly also told its readers that the "summit plateau contains about seventy acres of ground, is slightly rounded, and completely covered with large boulders." The magazine described the station itself as "a substantial one-story building with walls two feet thick and . . . erected at a cost of $2,500. The structure faces the east."[2] (In time, the signal station would be called Summit House and the cabin below at Boehmer's camp would be replaced in 1875 by a lodge called Lake House.)

Sergeant Boehmer apparently did more than give graceful speeches at the dedication for, nine days later, he was relieved of his duties for "gross mismanagement."[3] Some say he freely gave out food and drink to visitors, faint from the rigors of the climb. Some say he handed out hams and bushels of cornmeal to the construction workers at the station. Whatever he allegedly did or did not do, George Boehmer went back east, and, in 1876, joined the staff of the Smithsonian Institution on the Bureau of International Exchange desk.

General Myer put Sergeant Robert Seyboth in charge of the signal station for the first winter, with Privates Leander A. Lemman and James H. Smith assigned to assist him. Sergeant David H. Sackett and Private Edward W. Boutelle took over duties in the Colorado Springs office. For Pikes Peak duty, Signal Service sergeants earned $77.25 per month; privates were paid $60.43 per month. The new weather post had a wind vane, barometer, minimum and maximum thermometers, hygrometer, anemometer, nonrecording rain gauge, tool box, wall clock, and a multitude of federal forms.

Robert Seyboth, like Boehmer, was a native of Germany; by the age of twenty-one, he had immigrated to the United States and enlisted in the army in 1867. He studied meteorology at Fort Whipple, Virginia, put together a weather station at Wilmington, North Carolina, and then moved on to head the station at Mount Washington—each time moving to a higher altitude.

Almost all the army men, and visitors as well, complained of headache and dizziness when they first arrived at the summit. Some had violent and embarrassing bouts with vomiting. Said Seyboth, "I myself was affected in this manner during the first two days, but feel quite well now."[4]

Another temporary disadvantage of the new weather station was its distance from a natural water source. The men were obliged to fill loaded demijohns from a small creek located three-fourths of a mile below the peak. All changed with the storm of October 22, however, for the three-day snowfall, which lay about in great drifts due to a fifty-mile-per-hour wind, became the winter source of water. The men regularly collected the snow in a tea kettle and various cooking pots and put them on the hot stove.

On November 1 at 5:42 A.M., Sergeant Seyboth began daily weather reports and, on November 6, those notes began to flow down telegraph line no. 99 to Colorado Springs—when the line worked. The contractor that Colorado Springs and Western Union had selected to do the Pikes Peak work had used wire of inferior quality. Moreover, the signalmen soon learned that the lines could snap from heavy frost that built up on the wire to a diameter of six inches. They could also break from the weight of snow from avalanches, could go down under falling trees, or could stop service due to unexplained or unknown causes. Sometimes the wire looked like a "huge, white snake . . . twisting and writhing under the influence of a stiff breeze."[5] The trouble could occur anywhere from the base of the mountain up to the summit. Private Smith grumbled that he was the one most often sent to find the break in the line and repair it.

After the first week of operation, life on top of the mountain achieved a certain routine. The first observations of the day were taken at 5:30 A.M. and the last evening report was made at 9 P.M. James Smith remembers the first season on Pikes Peak:

> Winter set in with the utmost rigor almost the first of November, although even summer on the peak would have been called winter in other localities. Snowstorms were of almost daily occurrence, and the temperature fell steadily. By the tenth of November the thermometer touched zero, and from that time it was intensely cold.[6]

Private Smith did not like the mountain wind, which

> began to blow with great velocity, fifty, sixty, and seventy miles an hour. The cold and the snow were not so much to be dreaded, had it not been for the wind. When it was calm and clear, no matter how cold, we thought nothing of taking a stroll in our shirt sleeves; but when the wind blew we could not put on enough wraps to keep out the cold.[7]

Smith was not the only observer affected by Pikes Peak winds. One winter night, Sergeant Seyboth wrote in his log of the wind and likened the "howling and roaring" to a "mighty monster [coming] over the summit, as if seeking its prey."[8]

When James Smith had to venture outside, which was often because the telegraph lines snapped again and again, he described the scene about the station: "Of plant life, even in summer, there was none, not a shrub or a speck of moss, and in winter, when the entire mountain range was covered with snow, the scene was desolation itself."[9]

On November 2 at 4:45 in the afternoon, Sergeant Seyboth made a note of "a curious and remarkable phenomenon" that he and his men were to observe over and over:

> In the distance to the northeast a heavy mist was hanging over the plains. On this mist was depicted a perfect shadow or profile of our Peak by means of the setting sun. The image was very realis-

tic; so much so, that one could hardly credit that it was merely a shadow and not a lofty peak rising abruptly from the prairie. This continued for the space of fifteen minutes, when it gradually melted away in the mist as the sun was setting behind the gigantic peaks of the "Snowy Range." During the presence of the phenomenon we watched it with curious interest, and much regretted when it faded from our sight.[10]

On November 9 the station had its first visitors since the dedication. Three men who wanted to see a Pikes Peak sunrise had started for the summit at 2 A.M., but they arrived just fifteen minutes too late. Needless to say, Sergeant Seyboth reported, the visitors "were not very enthusiastic when they started for home again."[11]

The following day, a lady and her guide visited the station. Seyboth noted that she "expressed much surprise that a station should be established at such an out-the-way place, when there were plenty of better places down below! *Sancta simplicitas,*" sighed the sergeant, philosophically.[12]

Perhaps the lady would have had her opinion of the station location much reinforced if she had made her journey one week later on Monday, November 17. "This day will be remembered by us years hence, reported Seyboth. The temperature had risen to 32 degrees overnight. With the rise came winds that increased to "hurricane" intensity by midnight. The sergeant lay in his bed—"Sleep was out of the question"—listening to the storm outside.

> The northwest end of the roof, although weighed down by heavy rocks, was frequently lifted up several inches above the wall, tearing away the plastering, which fell in showers over my bed, desk, and the floor. . . . I could hear the tin being wrenched off the sheathing-boards, where we had nailed it with great pains and much confidence in the security.[13]

At dawn, Sergeant Seyboth went out to check the humidity on the hygrometer. As he rounded the corner of the station, the wind picked him up and tossed him unceremoniously into a ditch six feet deep. He realized the storm was going to get much worse and he and the men would have to hoist more heavy rocks to the roof if their quarters were to remain intact.

Private Lemman was the heaviest of the three; therefore, Seyboth stationed him on the roof with a rope to haul up the rocks that the ser-

ROCKS AND MORE ROCKS. Some held down the station roof during high winds. Not a picture of a stable building. (*Western History Department, Penrose Public Library, Colorado Springs.*)

geant and James Smith forced up with poles. "It was terrible work, for often, after getting a rock nearly to the roof, a sudden gust of wind would knock all three of us off our legs, and the rock would barely miss falling on Smith or myself. However, we succeeded in getting up about a ton of rock, which effectually stopped the lifting of the roof."[14]

By afternoon the wind brought with it heavy snow and sleet, which continued throughout the night. In the morning, not surprisingly, the telegraph line was down. Again, Private Smith was sent to investigate.

He made it down to timberline, then came back to report there were no breaks along the line. From Colorado Springs, Sergeant David Sackett started up the trail through the forest; he repaired nine breaks in the wire where dead trees had toppled in the wind, snapping the line.

On November 19 Edward Copley, the wood contractor, came up to the station and stated to Sergeant Seyboth he intended to petition Congress for a permit to build a hotel near the top of Pikes Peak. Even though many thought Copley was entertaining a construction nightmare, Seyboth thought it was a good idea. He was already tired of feeding, nursing, and entertaining the many visitors who came up the rough trail to the station. Let a commercial enterprise feed and house them, he said; leave us alone to do our work.*

Unlike James Smith and some of the other observers on top of the mountain, Sergeant Seyboth never seemed to get bored or upset. He was, of course, not the one to be sent outside to climb about the rocks and snow repairing telegraph lines. Each change in the weather and each sighting of an animal along the mountain trail was a source of amazement to Seyboth. He devised his own entertainment if there was nothing to do. On November 20, for example, the sergeant played a game of checkers by telegraph with W. W. Allen in Colorado Springs.**

The weather men were not prepared for the difficulties of simple meal preparation at high altitude. Private Lemman, on kitchen duty, cooked a pot of beans for more than ten hours; the men tried to eat them but they were still hard and inedible; they pushed their plates away. Forget beans. The Army's cheapest staple could not be cooked at that altitude. Lemman tried cooking them in what was called the Warren Cooking Apparatus with no success.*** Seyboth reported this curiosity over the telegraph, adding that potatoes had to be boiled for four hours or more before they were soft enough to eat. Water boiled at 178 degrees instead of the usual 212 degrees at sea level. When the men poured a cup of coffee, they had to drink it quickly before it turned cold.

Temperature and altitude continued to fascinate Robert Seyboth. He

---

*Two years later, Copley did build Lake House overlooking Moraine Lake, about four miles below the summit.

**Mr. Allen was the man who had written Washington about the need for a weather station on Pikes Peak.

***Precursor to the pressure cooker.

discovered that when he put a dishpan full of snow dug from an outside drift on the stove, the snow at the bottom melted and boiled even though the snow on top remained solid and several inches thick. "Thrusting my finger through the snow, I found that it had formed a solid crust above the boiling water; but even this was not sufficient to condense all the steam, which escaped with loud hisses through its icy confines."[15] Apparently, Seyboth shared with his boss, General Myer, an intense intellectual curiosity about life around him.

On November 22 the telegraph line broke again. Sergeant Seyboth sent out Private Smith as usual to find the problem and repair it. At 11 A.M., Smith found the break and fixed it; on the way back, however, he got caught in a snowstorm and lost the trail. For two hours, he crawled over the rocks, frostbitten and exhausted, until he found the door to the station. Seyboth described Smith's ordeal in the log and concluded, "Duty on Pikes Peak is not fun."[16]

But most of the time, it was fun for Sergeant Seyboth. His daily logs were filled with the minute details of temperature, wind velocity, and humidity, to be sure, but he also entered items unrelated to the weather. He was astonished at the unusual tameness and boldness of the resident mice:

> Not only are they running about the office in broad daylight and getting under my pillow at night, but this afternoon a very intelligent-looking mouse got actually on the desk while I was writing, and almost allowed itself to be touched. . . . Lemman threw it a few cheese-crumbs, which it devoured, sitting on its hind legs like a squirrel. These mice are of a light-brown color, having long ears and sharp, pointed noses, singularly expressive and large eyes, and are not nearly as repulsive looking as their brethren of the lower regions.[17]

Two days later, Seyboth noted that

> a strange-looking animal jumped on the cross-bar of our instrument-shelter and looked through the window, as if it wanted to come inside. It was nearly the size of a cat, with a long, bushy tail and mouse-colored pelt. Its head resembled that of a rat, but much larger. Before we could open the window the animal took fright and disappeared. I intend to set a trap for it.[18]*

---

*Seyboth would find out later that it was a marmot.

That same day, Sergeant Sackett excitedly telegraphed from Colorado Springs to Seyboth that, while camping beside Mystic Lake, which everyone thought was a barren tarn, he dipped a net into the water and caught a shrimp! Perhaps the discovery of the lake shrimp took hold in Seyboth's mind. The chief observer was beginning to realize that he was becoming a celebrity via the telegraph wire. His reports on beans and potatoes, storms and winds, mice and rats with bushy tails, reprinted in the local papers, were enthusiastically received by a growing audience. The public loved these tidbits and wanted more. He could not let them down. So he told them of fierce hundred-mile-per-hour gales and wild storms with hail as big as pumpkins. A series of thunderstorms, he told his readers, had left him deaf for several weeks.

Evidently, General Myer did not object to the sergeant's public relations efforts. Of course, the chief signal officer may not have known of Seyboth's little stories, as Myer was off to the First International Congress of Meteorologists in Vienna, where he explained his recently developed weather forecasting system. Whether or not Myer relied on the expertise of Cleveland Abbe and Increase Lapham, he must have learned a lot in a short time because, from all accounts, his remarks were well received by the scientists. By year's end, seventy-eight weather observer stations had been established throughout the United States. Some four hundred volunteer weather observers from the Smithsonian Institution were transferred the following year to the Signal Service, but, one by one, they dropped out of the project, stating they did not like to be attached to a military organization. Others in the scientific community felt the same way. A national weather service should be under civilian control.

Back at Pikes Peak, Sergeant Seyboth initiated a special program of his own, the first of many stories about the summit that were to entertain citizens for a number of years. Seyboth's literary offering, which was printed in the December 6, 1873, edition of the *Colorado Springs Weekly Gazette,* was an outrageous success.

There is, he reported, in understated but morbid tones, some sort of monster in the cold blue waters of Mystic Lake, located just below the peak's timberline. One day, he continued, while he and his faithful burro were following the trail beside the lake, he heard a loud splashing noise coming from the center of the water. He turned to study the lake. Something, a something that was at least a hundred feet long, was moving swiftly across the water. Its horrid body was an obnoxious pale brown and was covered with great scales. Its head, held up by a long, skinny neck, raised up six feet above the water. The monster's eyes were truly small for such a large animal and seemed to glance rapidly in every direction as the slimy creature glided through the water.

That was not the end of Robert Seyboth's monster story, however. He felt the need to embellish the tale further and give it a Western flavor. To that end, he wrote that he and his faithful burro fled down the Bear Creek Trail, where they met three Ute Indians who were hiding behind a boulder, shaking with terror. Had they seen the Mystic Lake monster, the signalman asked them? Seen it, they cried, we have just escaped from its terrible claws. They told the sergeant that the creature had "ruled the lake for centuries and was particularly fond of Utes, having eaten seven since March."[19]

It was a sea serpent, of course, and the news of Seyboth's "discovery" made headlines not only in the American papers but also in the European press. The news traveled to Scotland, where Loch Ness residents were highly outraged to learn that an American sergeant was encroaching on their sea serpent territory.

6.

A Thanksgiving gift. Private Smith has enough. Animal visitors.
The end of a patriarch. A sea in the Rockies. Christmas at the top.
Seyboth gets a shock. A curious "fire." January is terrible. An unusual snow.
A light to the north. Colors around the moon. April and May are worse.

THE MEN ON TOP OF THE mountain celebrated Thanksgiving Day
1873 quietly; there was no turkey, but James Smith baked a short-
cake and some dried-apple pies for dinner. Robert Seyboth entered in
the log a prayer that concluded: "I feel deeply that we three laborers in
the field of science have also just cause to be thankful for His protection
from the dangers of our position."[1] Nature seemed to be entertaining
the men, for the holiday featured a variety of unusual cloud formations:
a fiery mass appeared at dawn; later, black stratus clouds moved above
the men; a sea of cumulus clouds stretched across the prairie in the late
afternoon. In addition to telegraphed best wishes from Colorado Springs
residents, a thoughtful young lady gave Seyboth a female kitten, which
made the journey up the trail in Leander Lemman's overcoat pocket.
Once deposited on the floor, the kitten made a quick adjustment to life
at the summit and began to hunt for mice almost immediately. A number
of bushy-tailed "rats" had met their end in the sergeant's traps. A visitor from town studied the creatures and told Seyboth they were "rock-
martens," or marmots, a rodent species common to the Rocky Mountains
and sometimes called whistlepigs.

In the station log for the third of December, Sergeant Seyboth entered this ominous note: "This has been a terrible day."[2] James Smith
was sent out to look for broken telegraph wire below timberline the
previous afternoon and decided to stay the night at the lakeside cabin at
Boehmer's camp. In the morning, Smith looked up at the sky; the
weather seemed stable enough to start for the peak. As he climbed
higher and higher, however, the air turned colder and colder; his eyes

were blinded by the glare from bright snowdrifts. When the private approached the halfway mark between timberline and summit, a gale hit the mountain with tremendous force. His hands, his face, and his clothes were thick with layers of ice and snow.

The trail grew steeper and steeper. Smith crawled on his hands and knees toward the peak but could not avoid the stinging sleet and howling wind. Dragging himself between the boulders at the summit, he saw the station house through the snow and fog and made for the door. He was able to stagger through the door upright but immediately fell to the floor in a faint. Seyboth had not expected Smith to attempt the climb; as it was, the private must have been struggling through the snow for more than eight hours. Seyboth rubbed the victim's hands with snow but felt certain that Smith would lose parts of all his fingers. Luckily, the private did not lose his fingers but carried one of his hands in a sling for more than a year; each succeeding winter, his tingling hands reminded him of his Pikes Peak adventure. Smith's "sphere of usefulness" was over in the Signal Service;[3] soon he left the West and became a staff member of the Smithsonian Institution in Washington as George Boehmer had done before.

In addition to mice and marmots, the men were beginning to see signs of other animals around the peak. Leander Lemman observed fresh bear tracks on one of his journeys up to the station. He was glad it was only tracks that he saw and not the beast itself, as he was unarmed. Later, the men saw a shy animal that they called a mountain-bison. They met deer and mountain sheep on the trail. Sergeant Sackett was so excited about another capture on Mystic Lake, this time a water bug, that he again used the telegraph to relay the news to Seyboth, just as he had done with the shrimp. The would-be naturalist brought Seyboth a beautifully colored bird he had killed at timberline. The station head packed it carefully, together with one of the marmots he had trapped, and sent them off to Captain Henry Howgate in Washington for further study.

Seyboth himself was thrilled to spot near the station what he called a European sparrow, its feathers puffed up to ward off the cold. The men put out crumbs for the little snowbird, as they came to call it, which became a regular guest. One early morning, the sergeant saw a good-sized dark gray bird, larger than a blue jay, but, when he ran for his gun and came back outside, the creature had vanished in the fog.*

---

*The bird was probably a Clark's Nutcracker, a common high-country resident.

Late in January, Seyboth reported that he had found a large rat, a real packrat, not a marmot, "sunning himself on the door-sill without attempting to run" when he stepped out in the morning. The sergeant coaxed the creature into the office with the idea of studying the rat and taming it, but it died in a few hours, most probably of old age. "[H]is long ears showed signs of rough usage, as if he had been worsted in a family quarrel. . . . His fangs were fully three-eighths of an inch in length."[4]

On some days, the men could see virtually nothing from the mountaintop. On these occasions, the summit "was enveloped throughout in a dense cloud of frozen vapor, cutting off all view beyond the distance of a few yards." But Seyboth's view on December 20 was another matter:

> A most wonderful and novel sign met my astonished eyes on opening the door this morning; the plains had disappeared, and in their place there was, as natural as could be, the ocean, with rolling billows and spray dashing against the rocks of the mountain's base. . . . But the most striking effect was produced by the rigid immobility of the whole picture. . . . Imagine the ocean stirred up by a stiff breeze and then, by some magic, to be suddenly turned to ice, billows, spray and all . . . but far above us stretched forth an unbroken sheet of cirro-stratus clouds which obscured the sun, and prevented his rays from falling on the frozen sea below.[5]

The men watched the "ocean" for several hours until it lifted and joined the high clouds above, whereupon it started to snow. They learned later that this was not a unique weather event, for they saw it again and again.

Christmas 1873 started as another quiet holiday. However, a heavy gale began blowing just before sundown, causing the heavy boulders on top of the roof to "roll about like so many pebbles." "By keeping the stove at red-heat and sitting close to it, we managed to bake on one side and freeze on the other."[6] Even so, the men enjoyed Leander Lemman's holiday feast of roast venison, potatoes, and cranberry pie, followed by a whiskey toast.

Now that the men had been at the summit for several months, they felt they had enough experience to make tentative generalizations about the winds and their seasonal directions. They noted that most of the winter snowstorms occurred with a wind coming out of the southwest; when the gales changed direction, the snow would abate. If the winds again veered to blow from the southwest, snow began to fall again. Proof for Benjamin Franklin's earlier observations. Later on, in the spring, they would begin to see a different wind pattern.

Each time the observers watched a hailstorm, they logged the size of the hail and the duration of its fall. They examined the hailstones and noted several internal layers, as if they had bounced up and down through upper areas of snow and sleet. Most of the pellets were the size of a pea and not that of a pumpkin, as Seyboth reported to his fans. A later signal crew, however, found hailstones with diameters of one and two inches. Eager to know more about these snow pellets, they studied the accompanying weather patterns and reported to Washington their view that "in all hail storms, the fall of hail entirely ceases for about half a minute following a heavy electrical discharge, and . . . the hail-fall is considerably heavier for some little time following the discharge than before."[7] Sixty years later, meteorologists who studied the origin of hail may have used the notes of the Pikes Peak observers.

On November 23, early in the morning, Sergeant Seyboth attempted to fire up the stove and accidentally touched its pipe. "Immediately a bright and long spark passed with a crackling sound between it and my fingers." Ever curious, Seyboth touched the stovepipe again, but he could not raise another spark. He wondered why. By early spring, the sergeant was beginning to understand this electrical phenomenon and was used to sudden shocks, apparently, when he reported, "the stove played one of its old tricks again this morning." When he touched the stove, he found that it had "transformed itself into a reservoir of atmospheric electricity during the night; it maliciously transformed the nimble fluid to my fingers, with quite a severe shock."[8]

On another occasion, the sergeant found the telegraph relay in the office apparently working by itself. He moved to adjust it and fell to the floor from a violent shock. Despite the jolt, he picked himself up and looked out the window to check the weather and noted that both snow and sleet were falling.

> I cut out the instrument and screwed the lightning-arrester closer, and in a few minutes a continuous stream of electricity passed between the two plates of the arrester, with a loud noise, resembling that produced by a child's rattle. The electric fluid passed, not in sparks, but in five or six continuous streams of light as thick as a pencil-lead, for two or three minutes at a time, with short intervals between.[9]

A few nights later, when Seyboth was getting ready for bed, he touched his underwear and received two quick shocks. This seems to happen most often after a snowstorm, he decided, and entered the underwear event in the log.

Robert Seyboth was able to add another subject to his repertoire of unusual weather phenomena with which he entertained by telegraph his many fans in Colorado Springs and beyond. On a number of occasions, he said, a brilliant violet-and-white light had appeared on the cups of the anemometer, forming fiery rings as the instrument turned around. Light also appeared on the revolving wind vane; it looked like a flaming arrow swinging round and round. Sometimes, he reported, all of the telegraph line leading into the station was "outlined in brilliant light, which was thrown out from the wire in beautiful scintillations."[10] These little jets of flame constantly jumped about the line, first in one direction, and then in another.

When the men touched the light, they felt no heat or shock but only a slight tingle through their fingers. The little jets of fire would vanish with a popping, crackling sound. Or the light might transfer to their outspread hands, elongating their fingers with three-inch brilliant cones. One of the braver observers put his tongue to the telegraph wire when it was covered with the fire. He expected a shock but, to his intense surprise, he did not receive one, nor did his tongue get burned. A bearded visitor to the station during one of these brilliant displays was alarmed when sparks of light jumped about his hair and whiskers like a colony of fireflies. The pack animals were decorated with sparks "like a firebug" as snow alighted on their backs, then disappeared.[11]

This phenomenon was Saint Elmo's fire, and since it did not hurt, the men were thrilled when it appeared on the peak. So was the public, who loved to read about the luminous, feathery, violet-and-white jets—sometimes the cones were blue and green—that danced about the highest reaches of the summit. Robert Seyboth told his fascinated audience that occasionally the color display was accompanied by a singing sound or a crackling noise. This phenomenon could only occur during or just after a storm, "when the air is moist; the most favorable condition is during the time a light, soft snow is falling."[12] Sometimes the pungent smell of ozone burned the noses of the signalmen as they watched the display.*

---

*Saint Elmo's fire got its name from an early Italian bishop who became the patron saint of sailors. In the days of clipper ships, the appearance of the bright light dancing along mast and topsail was a welcome sign to the men on board and meant that a storm was nearly over. In Portugal, Saint Elmo's fire was called *corposant* (holy body) because the Portugese felt its appearance was due to unexplainable supernatural events. The bright fire can appear on other high

After their first experiences with Elmo's fire, the weathermen dared hope that, come spring, female visitors would arrive at the proper time to be entertained by the colored jets dancing along the telegraph wire and the bright light firing up the men's hair and whiskers like lanterns. What fun it would be to confound them.

January 1874 was a difficult month for the Signal Service men at the peak. Days were dark and gloomy, fierce winds raged unabated, and snow drifted through the cracks in the cabin, gathering everywhere in icy heaps. To make things worse, the men could glimpse from time to time Colorado Springs's warm, sunny plains below. On rare occasions, however, the situation did reverse and those below suffered disagreeable weather while the men above basked in the sunshine.

Both Sergeant Seyboth and Private Cyrus Cramer, James Smith's replacement, suffered the pain of frostbitten hands and snowblind eyes. Private Lemman got an unpleasant attack of neuralgia, which was described as pain in his nerve endings. Edward Boutelle was very ill with inflammatory rheumatism and Seyboth contracted diphtheria, a common disease of the 1870s. Surprisingly, the others did not get the illness. Some of the visitors to the station during this period of bad weather experienced violent bouts with "Pikes Peak fever"—headache, nausea, and vomiting—an early example of altitude sickness that some present-day hikers suffer on a first trip to high country.

One night in February it snowed sideways! Seyboth reported this unusual fifteen-minute storm:

> The sky appeared to be perfectly clear and the stars shone brightly, yet snow was falling fast in good-sized flakes; the wind had veered to the north, from which direction the snow-flakes were coming, and I was particularly struck by the fact that they did not appear to fall from above, but that their path made a sharp angle with the summit.[13]

Despite illness and foul weather, the men persevered, recording their observations both day and night. On March 7, at 9 P.M., Seyboth saw "a faint whitish light on the horizon in a north-northeast direction from the summit."[14] Its arch was indistinct and had no motion to it, nor did it

---

places—the steeple of a church, the wing of an airplane. It can also appear on the horns of cattle, the ears of horses, and can even run up and down exposed blades of grass or along the top of a brass bedpost.

have streamers or "merry dancers" attached. Even so, the sergeant realized it was an aurora and surmised that Pikes Peak was too far south for possible observation of clear "northern lights." (These lights are more properly called aurora borealis, a seasonal electrical phenomenon that is most clear in arctic regions.) In June, however, the men did see another aurora, this time with some streamers attached, and, in later years, one signalman on the peak described an aurora as looking like the illumination from a far-off forest fire.

A rainbow around the moon greeted the men on the evening of March 23. Seyboth described it with characteristic grace: "As the fleecy clouds passed swiftly across the moon, they produced, successively, coronae of exceeding beauty, often exhibiting three different series of colored circles."[15] If the sergeant was correct in identifying the sight as a corona, he must have seen red in the outer rings.

Lunar halos were also sighted several times during the early spring. These phenomena are described as "colored or whitish rings and arcs about the . . . moon, when seen through an ice crystal cloud or a sky filled with ice crystals."[16] These halos were colored in the reverse manner from coronas, with the red rings nearest the moon. At first, Sergeant Seyboth theorized that the halos signaled a bout of bad weather. However, during a clear period the following June, he saw more lunar halos and discarded his theory.

April's heavy snow buildup at the summit was so severe that the sergeant predicted it could not possibly melt by late summer. Seven feet of snow now drifted about the weather station, sometimes up to the roof. In fact, he reported, "April has proved itself the most disagreeable month since the opening of this station in November last."[17] Difficult weather occurred during the winter months from winds blowing in from the southwest, but now, in April, the winds brought storms from the northeast and the southeast. May was no better, but the men now stoically regarded themselves as veterans of howling winds, heavy snows, hail the size of large peas, and sizzles and electric shocks from the telegraph wire, the stove, and the lightning arrester. All were well seasoned but Private Daniel O'Leary, one of the assistants, who became so unstrung during one of the electrical displays that he had to be treated for mental problems down in Colorado Springs; he was eventually declared unfit for any duty and never recovered his sanity.

Despite the violence of the weather, the men ventured outside from time to time. Once in a snowdrift the sergeant found a stunned ladybug, which he brought in to preserve for scientific study. Even though heavy storms continued, Sergeant Seyboth wrote in the log that he felt summer on Pikes Peak was definitely on the way—because the men now heard occasional claps of thunder, with or without accompanying lightning strikes—sounds of conventional summer thunderstorms.

On May 24 the sergeant went outside for a 2 P.M. observation and experienced a phenomenon he could not fully explain: "I heard the snow crackle . . . and at the same time I felt on both temples, directly below the brass buttons on my cap, a pain as if from a slight burn; putting up my hands, there was a sharp crack, and all pain had disappeared."[18] Obviously, he had encountered an electrical field, but one without the usual weather events that accompanied such charges he had come to expect.

The most violent storm experienced on the peak arrived at the end of May—thunderclaps, periods of snow and heavy sleet. The smoke produced by fierce discharges of electricity from the lightning arrester so terrified the men that they retreated to the kitchen, ready to fling open the door and make a run for it down the trail. The storm quieted down, and "passed slowly to the westward. . . . We were beginning to breathe more freely, when the wind veered to the southwest and brought back the storm in all its fury." More snow, sleet, thunder, and smoke from the lightning arrester. "The rattling of torrents of sleet, mingled with the incessant rolls of thunder, the blinding flashes and loud reports in the room, were enough to make the stoutest heart quail."[19] The men were particularly disturbed when they learned later that it was a sunny ninety-three degrees in Colorado Springs. All survived; doubtless some of the men wondered if perhaps this was the time to seek quieter employment elsewhere.

7.

THE SIGNALMEN AT THE station were beginning to regard themselves as experts in meteorology. By the middle of the summer of 1874, they were able to describe a curious weather pattern that they predicted could lead to flash flooding along the Front Range:

> The lower strata of air become powerfully heated, and are proba-
> bly, at this season of the year, when the surrounding mountains
> discharge their melting snows into the parks, heavily charged with
> moisture. The cold, heavy west winds descending the eastern slope
> of the main range, and wedging under the heated, moist lower
> strata, might explain the frequency of local storms on the peak
> during the hottest part of the day.[1]

It is not unusual for flash floods to course down Rocky Mountain streams already swollen with snow melt. Boulder, Colorado, suffered extensive damage from a one-hundred-year deluge in 1894.* Boulder was hit again with a twenty-five-year flood in early May 1969. Denver was badly flooded in June 1965. Rapid City, South Dakota, was subjected to an immense one-thousand-year flood in 1972. In late July 1976, even though mountain snow melt had already occurred, the Big Thompson River, west of Loveland, Colorado, boiled over into a flash flood, causing the deaths of 139 persons.[2] Perhaps the 1874 speculations by the

---

*Such a flood has a 1 percent chance of occurring in any given year at a particular location. A twenty-five-year flood has a 4 percent chance of occurring in a given year at a particular place.

THE FIRST HALFWAY HOUSE, built on the site of Boehmer's camp, was a relatively simple structure. (*Photo circa 1885; Starsmore Center for Local History, Pioneers Museum, Colorado Springs.*)

Pikes Peak signalmen on weather conditions that might lead to flash flooding proved valuable to later flood researchers.

The summer of 1874 was one of changes. Lieutenant A. W. Greely came out to attend to a number of things. First, he met with a grand jury to determine who had been stealing army goods since construction of the station the year before. Early in June the lieutenant arrived to inspect the summit office and determined the building was not suitable for human habitation—leaking roof, poorly fitting doors and windows, no place for storage of food, mortar crumbling in the walls. He recommended additions and improvements to the station as well as to the telegraph line. He further ordered a change in personnel at the summit, commenting that most of the men weren't worth much to begin with and that service at Pikes Peak caused health problems in even the strongest. "The question of influence of extreme elevation on health appears to be an unsolved one among the medical profession, and under such circumstances it does not seem proper to keep men on such a station over a year."[3] Private O'Leary was still unstrung; another assistant suffered from syphilis, the severity of which rendered him unfit for

VISITORS CLIMBED TO THE station's rickety roof to pose for their portrait, circa 1876. *(Starsmore Center for Local History, Pioneers Museum, Colorado Springs.)*

mountain duty. Only Sergeant Seyboth escaped the wrath of Lieutenant Greely and received good marks. When Greely concluded he had everyone behaving himself, he left for Washington. Meanwhile, wood-gatherer Ed Copley was making good on his promise to build a hotel on Mystic Lake below the summit; by August he was ready for his first guests. Even Mystic Lake, which fronted the new Lake House, was given a change of name to Lake Moraine. Copley was also given a $700 contract by Lieutenant Greely to build the addition to the original building, which he completed by early fall.

Sergeant Seyboth continued to be both irritated and intrigued with the growing numbers of official and unofficial visitors to the weather station. Despite the extra work, however, the men enjoyed studying the tourists and their varied reactions to the view at the summit. These adventurers arrived in various states of health. Often, the sergeant's men were obliged to minister to those who became ill and vomited from the rigor of the climb and the abrupt change in altitude. "A free use of whiskey as a stimulant in climbing is a fruitful source of indisposition," reported Seyboth philosophically. A strong cup of tea led to the recovery of most and was acknowledged by visitors to be "the best thing they had tasted for a long time"—a specific for Pikes Peak fever.[4]

Of course, many did not make it to the top of the mountain at all and turned back to Colorado Springs, disgusted with themselves. On the other hand, a small boy who arrived with a party in June, Seyboth said, "stood the trip like a regular." A sturdy climber with only one leg made the summit. Another group included "two rough-looking customers, who vented their feelings in a savage whoop as soon as they reached the summit. I thought 'Injuns sure' and looked after carbine and revolver, but they turned out to be a well-disposed Scotchman and a taciturn Missourian, both laborers, apparently."[5]

On June 21 Seyboth and Lemman were shouted out of bed at 4:30 A.M. by seven rowdy men who had come up to see the sunrise and were irritated that they had missed the event by a quarter of an hour. The seven hikers went down the trail at 7:30 A.M., a little quieter. Others would arrive in the middle of the night, and, not wishing to disturb the sleeping signalmen, politely stayed outside shivering until dawn. Numbed by the morning cold, they no longer cared if they saw the sunrise or not. All they wanted to do was to enter the cabin and "hug the stove."[6] Even those who made their way into the warmth of the station often missed the first appearance of the sun because of cloudy or rainy weather.

One day in late June, fourteen visitors arrived, many of whom were Texans, noted Seyboth, disparagingly. Ministers, photographers, surveyors, editors, scientists from Sweden and Germany—ninety-one in all—made it to the summit in June—some fit and some fragile. Of great sadness to the signalmen was that "no lady visitor has gladdened our hearts as yet."[7] Evidently, they had forgotten their female visitor the previous November, who announced that the top of a mountain was a silly place for a weather station.

The signalmen were obliged to cook for some of the tourists as well. Food supplies that were meant for three or four army men were stretched to feed any number of uninvited guests. Some brought welcome gifts of butter and sugar. One group of tourists, graciously invited to supper by the weathermen, looked at the modest spread of "beans, tea, and bread and butter," and disdainfully declined.[8] (Sergeant Seyboth had previously reported the men could not successfully cook the army-issued beans at a 14,000-foot altitude. Evidently, they had finally learned the art of high-mountain bean cookery.)

Despite these distractions, Sergeant Seyboth was able to concentrate on the study of a "great many bugs, spiders, and flies, also a large bumblebee. . . . All of these insects I found more or less benumbed on the snow." A good-sized butterfly entertained the men one June day as it danced about the summit. Shortly thereafter, Seyboth was delighted to find the first grasshopper of the season. With the advent of regular summer thunderstorms also came mosquitoes; their appearance truly surprised him as he had not expected to "see these little pests on Pikes Peak."[9]

When he was still at the station, former weatherman James Smith commented on the lack of any plant life near the summit. But that was winter. June produced an array of delicate and colorful flowers that hugged the rocks in what little soil there was. As the season grew warmer, however, the little flowers curled up and disappeared. With the dry spells came forest fires, and the men could see and telegraph to Colorado Springs the location of each ominous display of smoke in the area.

In July the first female visitor of the season to the Pikes Peak weather station was memorialized in Sergeant Seyboth's daily log—Miss Penelope E. Ambler of Mount Pleasant, Iowa. The men were overjoyed when the next group of lady guests arrived just in time for an afternoon thunderstorm that featured hail along with lightning. The observers were able to soothe away feminine terror from "sharp flashes and reports [which] came through the lightning-arrester." The hail crackled with electricity and Sergeant Seyboth was able to amuse and confound the ladies when his whiskers became electrified and gave off "audible hissing sounds." His scalp "prickled with hundreds of red-hot needles, and a burning sensation was felt on face and hands."[10] A visiting dog was also affected and howled piteously until it was let into the station. One male visitor reported that his hair was so stiff with electricity that his hat was lifted off his head. Even nearby rocks clattered and chattered. Doubtless, the ladies never forgot their singular afternoon on Pikes Peak. And the sergeant put forth his view that one of these strikes might well burn down the weather station itself.

Seldom did Robert Seyboth have any trouble expressing himself in the log; on the last day of July, however, he observed an optical phenomenon to which he felt he could not do justice with mere words—but he tried:

> A light rain was falling on the summit, while in the west the sun was just about to disappear behind the Snowy Range in an almost clear horizon. To the east from foot-hills north and south a magnificent rainbow spanned an arc of fully 270 degrees or three-fourths of a circle. White, fleecy clouds passed inside the arc from south to north, and on them was depicted a very accurate shadow

THE PALSGROVE BROTHERS packed wood for the winter up the trail, more or less regularly. (*Western History Department, Penrose Public Library, Colorado Springs.*)

of the house, with chimneys and parapet complete. Had one of us been on the roof, his figure and motions would have been faithfully copied on these clouds.[11]

During July 1874, 290 visitors came to see the sights, including fifty ladies, a fact that Seyboth was pleased to note in the log. Alas, he added, only four chairs could be found to accommodate them. No doubt he thought of what effect these reports would have on General Myer, so he added gruffly, "To make observations or to do any kind of work while such an invasion lasts is one of the most difficult and unpleasant tasks I have ever been called upon to undertake." Seyboth posted signs below stating that no food or accommodations were available at the top, but still the tourists came, some making "black-guards of themselves because they could not find Hotel accommodations at the Summit."[12]

August was characterized by a variety of events—thunderstorms, snowstorms, heavy smoke from a prairie fire, and 340 visitors. Two eagles landed outside the house, but fled when one of the men approached with a gun. A little ferret took up residence outside the station, enjoying a feast of coneys, rats, and mice.* The ferret seemed quite tame and ventured into the office when there were no visitors about, where it allowed itself to be handled and petted by one of the men without biting him.

A relatively quiet September allowed the men to catch up with their meteorological chores. Even the telegraph wire cooperated and seldom had to be repaired. The station itself was enlarged to better accommodate both men and supplies. The weather observation equipment was augmented to include "three standard barometers, one standard thermometer, two standard minimum thermometers, two standard maximum thermometers, three standard hygrometers, four standard anemometers, one standard rain gauge, one small wind-vane."[13]

But Sergeant Seyboth decided that he had had enough of this arduous life; his health had deteriorated due to his earlier bout with diphtheria and to repeated freezing of fingers and toes. His weight had dropped twenty pounds. He asked for reassignment to an easier post. General Myer noted that the sergeant had "discharged his duties at this isolated station in the most satisfactory manner" and granted his request.[14] Seyboth transferred to Philadelphia for a time, then back to his original posting at Wilmington, North Carolina.

---

*The coney is the Rocky Mountain pika—a small, tailless mammal related to the rabbit.

8.

**An eagle on high. John Brown's first storm. Changes in the staff.
O'Keeffe meets the judge. The terrible rats.
A gravestone for Erin. O'Keeffe goes east.**

A S EACH SUCCESSIVE GROUP of signalmen started duty on the summit, they read the logs of earlier observers and learned, more or less, what to expect. How different it was, however, to read about a sea of clouds, unusual animals, thunderclaps and lightning strikes, and telegraph wires and whiskers that lit up than to experience these phenommena firsthand.

Sergeant William Theodovious, who replaced Robert Seyboth in September 1874, was awed by the flight of an eagle some three thousand feet above the summit. He had observed eagles on the peak itself but to see a bird soaring at that height was extraordinary. Evidently, the observation of eagles was not enough for Washington. Sergeant Theodovious was relieved from his post the following February 1875 as "unfit for duty."[1] What he did or did not do is lost to history although it was well known that the sergeant enjoyed loosening large boulders with a crowbar and sending them crashing into the crater near the peak. During the same winter, in an apparent governmental clean sweep, three observer assistants were also removed from the station for misconduct, the nature of which was not explained either, at least not in print. As Theodovious departed from Pikes Peak, twenty-five-year-old Sergeant John V. Brown took charge.

Brown expressed himself with some force in a report of his first experience with a violent storm on July 5, 1875:

> Shortly before noon a terrible electric storm began. At first its effects were felt or seen on the line, but at about 2 P.M. its presence was felt everywhere on the summit. From the lightning-arrester

there was a constant stream of flame, filling the building with sul-
phurous smoke and rendering it almost unbearable; out of doors it
was even worse. There was a constant cracking noise in the air, as
though made by small pistols.[2]

Not a bad report for a signalman not endowed with Robert Seyboth's
imagination. Brown added that his pulse just before the storm was 76
beats per minute; when the storm broke, he took his pulse again and
found his heart was pounding at 120 beats per minute.

Two new assistants had joined the staff to replace the three men who
had been sacked. Private Thomas B. Bowlus reported for duty after the
first of the year. Another private named O'Keeffe was also ordered to
assist Sergeant Brown. Both men were given satisfactory marks by the
incoming sergeant, who sent favorable progress reports to Washington.
Within a few years, however, General Myer may have reflected on
what he had done in earlier days to deserve the likes of John Timothy
O'Keeffe.

The first time Private John O'Keeffe came to public attention was
during his off-hours in notorious Colorado City. The gray-eyed, black-
haired signalman was refreshing himself in a number of saloons in the
company of one Cherokee Charlie. Charlie insisted that John Timothy
O'Keeffe was an Irishman; the private insisted just as strongly that he
was an Italian. Neither man would give up his point of view and more
than spirited words ensued. The barhoppers' mistake was to move the
discussion into pure and pristine Colorado Springs, where they fought
until both men became "much bruised and mangled." The town consta-
ble was called in to separate the fighters; he brought them before the
bench of Justice of the Peace Eliphalet Price, whereupon the barflies
were "cast into the town bastille" as they had insufficient funds for
bail.[3]

Although this first meeting between Judge Price and Private O'Keeffe
was of a judicial-criminal nature, the event marked the beginning of a
firm friendship; the men recognized kindred interests in one another.
They became regulars in the saloons of wicked Colorado City and,
while in their cups, began to reflect on the possibilities of writing a tall
tale to entertain the public.

In truth, O'Keeffe was not of Italian descent but was an Irish-
American born in New York City. As a young boy, he dreamed of becoming a naval officer. When he reached the appropriate age, he applied
to the U.S. Navel Academy at Annapolis, handily passed the entrance

MASTER STORYTELLER John Timothy O'Keeffe. (*Denver Daily News,* July 25, 1895).

exams, and was accepted as a midshipman. O'Keeffe's reputation as a teller of wild tales had already developed by that time; apparently, his excessive imagination was too much for the officers at Annapolis, despite the fact he was neither vicious or immoral. He was saved from outright expulsion, however, by a New York congressman who liked the young man and thought he might do well in the expanding Signal Service. So, instead of following his childhood dream of training for a life at sea, young O'Keeffe signed up for a five-year stint with the military in May 1874 and found himself at the top of a western mountain peak on January 5, 1875.

O'Keeffe's drinking companion, Eliphalet Price, was both a lawyer and an aspiring journalist. Born in New Jersey, he had come to Colorado Springs in 1872 and founded the weekly *Free Press*. According to the local papers, sixty-one-year-old Judge Price "took part in many prominent affairs of public enterprise," and "was one of the most respected citizens in our town." His sedate and judicial manner belied his avid interest in the bottle. In short, Eliphalet Price "seemed to love printer's ink better than the bar, and the saloons better than the bench."[4]

Off-duty hours found Private O'Keeffe and his new friend Price, as well as another drinking companion, an unnamed British newspaper editor, in Colorado City saloons, where they joined forces in creative writing projects. Their first public production was overwhelmingly successful and was reprinted not only in the Denver papers, but also in the press of England, Scotland, Ireland, France, Austria, Russia, Turkey, and Egypt. The original tale began:

> The vast number of rats inhabiting the rocky crevices and cavernous passages at the summit of Pike's Peak, Colorado, have recently become formidable and dangerous. These animals are known to feed upon a saccharine gum that percolates through the pores of the rocks, apparently upheaved by that volcanic action which at irregular intervals of a few days gives to the mountain crest that vibratory motion which has been detected by the instruments used in the office of the United States Signal Service.
>
> Since the establishment of the station, at an altitude of nearly 15,000 feet, these animals have acquired a voracious appetite for raw meat, the scent of which seems to impart to them a ferocity rivaling the starved Siberian wolf. The most singular trait in the character of these animals is that they are never seen in the daylight. When the moon pours down her queenly light upon the summit, they are visible in countless numbers, hopping among the rocky boulders that crown this barren waste, and during the summer months they may be seen swimming and sporting in the waters of the lake, a short distance from the crest of the Peak, and on a dark, cloudy night their trail in the water exhibits a glowing, sparkling light, giving to the waters of the lake a flickering, silvery appearance.
>
> A few days since, Mr. John O'Keeffe, one of the government operators at the signal station, returned to his post from Colorado

Springs, taking with him a quarter of beef. It being late in the afternoon, his colleague, Mr. Hobbs, immediately left with the pack animal for the Springs.

Soon after dark, while Mr. O'Keeffe was engaged in the office, forwarding night dispatches to Washington, he was startled by a loud scream from Mrs. O'Keeffe, who had retired for the night in an adjoining bedroom, and who came rushing into the office screaming: 'The rats! The rats!'

Mr. O'Keeffe with great presence of mind, immediately girdled his wife with a scroll of zinc plating, such as had been used in the roofing of the station, which prevented the animals from climbing upon her person, and although his own person was almost literally covered with them, he succeeded in encasing his legs each in a joint of stovepipe, when he commenced a fierce and desperate struggle for his life with a heavy war club preserved at the station among other Indian relics captured at the battle of Sand Creek.

Notwithstanding, hundreds were destroyed on every side they seemed to pour (with increasing numbers) from the bedroom, the door of which had been left open. The entire quarter of beef was eaten in less than five minutes, which seemed only to sharpen their appetite for an attack on Mrs. O'Keeffe, whose face, hands and neck were terribly lacerated.

In the midst of the warfare, Mrs. O'Keeffe managed to reach a coil of electric wire hanging near the battery, and being a mountain girl, familiar with the throwing of a lariat, she hurled it through the air causing it to encircle her husband, and spring out from its loosened fastings, making innumerable spiral traps, along which she poured the electric fluid from the heavily charged battery. In a moment the room was ablaze with electric light and whenever the rats came in contact with the wire they were hurled to an almost instant death.

The appearance of daylight, made such by the coruscation of the heavily charged wire, caused the rats to take refuge among the crevices and caverns of the mountains, by way of the bedroom window, through which they had forced their way. But the saddest part of this night attack upon the Peak is the destroying of their infant child, which Mrs. O'Keeffe thought she had made secure by a heavy covering of bed clothing, but the rats had found their way to the infant (only two months old) and had left nothing of it but the peeled and mumbled skull.

A late bulletin ended the story: "Drs. Horn and Anderson have just returned from the peak. It was first thought that the left arm of

JUDGE ELIPHALET PRICE, writer of tall tales. (*Starsmore Center for Local History, Pioneers Museum, Colorado Springs.*)

. . . O'Keeffe would have to be amputated, but they now believe it can be saved."[5]

Such excitement. Shortly after this item reached the European papers, a German scientist held a press conference, telling the newsmen that he had thoroughly researched the story. "Gentlemen," he said, "it is impossible for rodents to live at an elevation of 14,000 feet above sea-level."[6] Almost everyone who had hiked to the summit had seen a few rats, their presence, doubtless, of great inspiration to the storytellers. Although some spoilsports pointed out that signalmen were not allowed on active duty when they were married, that John O'Keeffe was not married, had never been married, and certainly did not have a baby girl, most Colorado Springs residents defended the story, saying all was possible. As the *Rocky Mountain News* stated later, "People are more ready to believe a good story than a dry truth."[7]

Citizen Eliphalet Price—regrettably, he had not won reelection as justice of the peace—"admitted modestly that his little opus had put Pikes Peak—and Colorado Springs with it—on the map, journalistically speaking."[8] He now signed his written efforts with the pen name Mucilage.

Since the opening of the weather station, the number of tourists had been increasing with each month of operation. When the rat story was published, the number of visitors to the summit rose dramatically. They came to give their respects to "mourner" O'Keeffe, who had quickly erected a gravestone over the burial of one of the Signal Service burros, on which was inscribed the poignant and dramatic verse:

> Fair Cynthia with her starry train
> Shall miss thee in thy silent rest
> And waft one sweet, one spheric strain
> To Erin dear among the blest.

> Erected by John and Norah O'Keefe, to the memory of their infant daughter, Erin O'Keefe, who was destroyed by mountain rats, May 25, 1876.[9]

No matter that a Springs resident had given a cat named Erin to the sergeant a short time before. Tourists, of course, wanted to see a picture of the O'Keeffe baby. Always ready to oblige, the sergeant had a number of baby prints made in town and sold them at the station for fifty cents apiece.

The Pikes Peak area suffered a plague of grasshoppers that summer;

MOURNERS FOR "ERIN O'KEEFE." John O'Keeffe may be the man standing third from right. (*Western History Department, Penrose Public Library, Colorado Springs.*)

one wonders why Judge Price and O'Keeffe overlooked the insects as a possible subject for creative writing.

How did Chief Signal Officer A. J. Myer react to all of the excitement? There is no way he could have missed the story. Perhaps O'Keeffe told him that he himself had nothing to do with the tale, that Judge Price was wholly responsible. Whatever were the general's private thoughts, O'Keeffe remained at the weather station filing his dry reports to Washington. When he was off duty, however, he and Judge Price continued to frequent Colorado City saloons together and began to ponder the subject for their next offering. They were aware, moreover, that they had great competition in circus man P. T. Barnum, who had recently made the papers with his story of the Great Muldoon, a mummified giant complete with long tail that had been "dug" from a site southwest of Pueblo, Colorado.

Sometime in the fall of 1876, John O'Keeffe was relieved from Pikes Peak duty by General Myer and sent to Fort Whipple for further training in order to be promoted to sergeant. What was General Myer thinking of? Did he dare hope that promotion might dilute the creative juices of signalman O'Keeffe? When he finished his classes, the new sergeant moved on to Philadelphia to help with the centennial exposition. Colorado Springs residents had to make do with the more mundane reports of the sergeant's colleagues until his triumphant return to Pikes Peak duty in 1880. Meanwhile, Judge Price continued to frequent Colorado City saloons and to develop stories about other weather station adventures.

9.

**More rings around the moon. A white visitor. Zodiacal lights.
The singing of crickets. Restless hair. A long blow.
Professor Abbe and the eclipse. A new planet.
Lieutenant Buchanan inspects. O'Keeffe is back.
Seven hundred deer and six mountain lions.**

S ERGEANT CHARLES M. HOBBS, who had succeeded Sergeant Brown
in January 1876, did not seem unduly sad at the departure of John
O'Keeffe, whom he found erratic and difficult to work with. Hobbs
added his own artistic touches to the official weather station log, however, with a discussion of some colorful rings that he observed around
the moon. The new summit observer studied the Signal Service manual
and decided the rings could not be characterized as either a corona or a
halo. A halo, he knew, had reddish arcs nearest the sun or the moon. A
corona also had red rings but they were at the edge of the effect, away
from the sun or the moon. Hobbs observed a phenomenon without any red
rings. Nearest the moon, he saw "bright golden color; this merging into
one of greenish tinge, followed by one of a distinct blue and then by yellow."[1] Alas, he was never able to put a name to these rings.

One morning, Sergeant Hobbs spied a "curious little animal" in the
woodshed. He noted in the log: "Have never heard of anything like it
having been seen here before." He described the creature to a long-time
Colorado Springs resident: "It was about six inches in length, had a
long, slender, and graceful looking body, a long tail, and was as white as
freshly fallen snow."[2] The experienced local told the excited Hobbs
that he had seen an ermine on Pikes Peak.

On February 7 of the new year, the sergeant had a hiking experience
reminiscent of James Smith in earlier days:

> I left the lake at 8:30 A.M. and arrived on the Summit at 2:30 P.M.,
> thus consuming six hours on the trip, which was one of the worst I

have ever experienced. It is useless to report descriptions of the terrible experiences which we residents of the Summit are at times obliged to undergo, for no description can give a just idea of the terrible reality. I have never had to force a passage through deeper snow than that which is on the trail now for as it is freshly fallen I would sink to the bottom at every step and with the wind enveloping me in dense clouds of drifting snow it was all but impossible to make any progress. I earnestly hope and pray that I will not meet with more than half a dozen more such experiences before I am relieved from here.[3]

A special study of pale yellow zodiacal lights, ordered by Cleveland Abbe from Washington, occupied Sergeant Hobbs's attention for several days during February and March 1877.* "The phenomenon is caused by the scattering of sunlight from a cloud of particles lying in the ecliptic."[4] The lights were quite clear in the east at sunrise, he reported to Abbe—the upper section clear and distinct, the lower somewhat fuzzy in appearance. The atmosphere surrounding the glow was a light purple. During the evening, Hobbs was rewarded with views of brightly colored coronas circling the moon.

Charles Hobbs was somewhat disconcerted one May evening just before supper when he stepped outside to hear the "singing of crickets" at two or three locations along the telegraph line.[5] When he approached the wire, the singing stopped; when he stepped away, the noise started up again. He did not offer a theory as a cause for this event, at least not in the log.

One June afternoon, Sergeant Hobbs thought he heard "a queer hissing" coming from the telegraph line outside the station. Again, he stepped out to investigate this new sound and found the noise came as well from the wind vane and a post set in a snow bank. Hobbs stood in several feet of snow. He could hear a few thunderclaps far away; a light sleet brushed his face. Suddenly the noise moved to the top of his head and "his hair became restless, and he felt a strange creeping sensation all over his body."[6] His nose burned with the smell of ozone. Hobbs made for the door of the station as fast as he could. When his feet touched the dry wooden floor, the strange feelings disappeared and he felt fine again. But now the telegraph key was acting up. Without thinking,

---

*Zodiacal lights are described as a diffuse glow seen in the west after twilight and in the east before dawn.

Hobbs put both hands on the apparatus and completed the circuit. The ensuing shock threw the sergeant to the floor with some force.

A few weeks later, as Hobbs stepped out into a summer snowstorm, he received similar sensations. His hair stood out from his head and "sang" again. The sergeant must have had a touch of the poet in him for he likened the noise to that of "burning evergreens." The equipment inside, he said, "produced a noise like that of swarming bees."[7] But the weather observers endured more than shocks, hisses, and restless hair that summer. A seventy-one-mile-per-hour wind blew across the station for more than twenty-four hours without abating. As each hour of howling and roaring went by, the men grew more and more apprehensive. Would the old roof hold?

July 29, 1878, was to be a very special day for Professor Cleveland Abbe, who still worked for the Signal Service although he had little to do with forecasting now and was more involved with special research projects. Abbe had come out from Washington one week early to prepare for the observation of a solar eclipse. More than a hundred scientists were to visit the summit; Chief Signal Officer Myer was also on hand. It took twelve pack mules, each making five trips, to carry the special equipment needed to study the eclipse. Abbe busied himself with arrangements, getting the instruments set up, tuned, and ready to go. But the scientist did not feel well and, when the great day arrived, Professor Abbe was not there to share in the excitement.

Evidently, Abbe did not adjust at all to the high altitude and developed the famous Pikes Peak fever. When someone made the suggestion that the professor move down to a safer altitude, Abbe brushed it aside, saying he would continue working. He became ill enough, however, that he was forced to change his mind and had to be taken down to Lake House on a stretcher. The week was not a total loss for the noted meteorologist, however. He was able, fortunately, without the aid of observing equipment, to watch at least part of the eclipse from this halfway point. He was astonished at the "great extent of the corona, larger than had ever been observed before in any eclipse." Abbe was particularly impressed with the streamers flowing from the corona. One, he said, extended "fully four degrees from the edge of the sun."[8]

Another visitor on that July day was astronomer Lewis Swift, who later reported that, aided by the clear atmosphere of Pikes Peak, he had seen a new planet. He called his "discovery" Vulcan—a name others have also used through the years for a "new" planet. However, further scientific testing to verify the existence of Professor Swift's Vulcan produced no further evidence of a new planet.

A MOONLIGHT VIEW OF PIKES PEAK, taken, no doubt, to encourage the tourist trade. *(Photo by W. E. Hook; Starsmore Center for Local History, Pioneers Museum, Colorado Springs.)*

The U.S. Signal Service seemed to be on a firm foundation by 1878. General Myer no longer had to worry about whether or not Congress would disband the army group, particularly the Weather Bureau personnel, as the observers were now called. In September of that year, *Leslie's Illustrated Weekly* had this to say about the general and the Weather Bureau:

> He is ambitious, but his ambition contemplates no merely selfish ends, and his aim is the broad and high one of serving, in the largest possible way, his countrymen and his age. . . . The Signal Corps* is one of the triumphs of our century, destined . . . to be instrumental in anticipating and averting disaster in all parts of the country.[9]

Despite these accolades at the national level, all was not well at the Colorado mountain station. Toward the end of its first decade of operation, Washington sent Lieutenant Buchanan to inspect the premises. The officer found the outpost in disarray. No new instruments for more sophisticated weather prediction had been installed. The signal quarters, Buchanan found, were in a "very bad and unfavorable condition."[10]

---

*The Signal Service was called the Signal Corps by some at this time.

A number of station personnel had been discharged for dereliction of duty. It appeared that the only reports being telegraphed from the summit were predictions of daily sunset, although someone during that period recorded a daytime temperature of sixty-four degrees on July 19, 1879, the highest logged since the station opened.

Even though Lieutenant Buchanan forwarded his dismal findings to Washington, no repairs or improvements were made at Summit House until 1879, when telegraph service was improved, two outside station walls were strengthened and rebuilt, windows were made more secure, and cans of fresh paint were packed up by mule to spruce up both office and kitchen.

The news from the weather station was brighter in January 1880, however. The new sergeant in charge sent this message to Washington:

> The citizens of Colorado Springs manifest a deep interest in the service, and are always ready and willing to promote its interests. Tourists, prospectors, cattlemen, and ranchmen generally are the parties directly benefited by the weather reports, and all of them testify to their general usefulness. The service at this point is highly appreciated, and is increasing in importance every year.[11]

The report had a familiar style to it, indeed. Sergeant John Timothy O'Keeffe had returned from a four-year stint at the Philadelphia Exposition to resume his Signal Corps duties. He had any number of new story ideas as well for his old friend Eliphalet Price, calculated to outdo any outpouring from their greatest competitor, P. T. Barnum. On one of his first ascents to the peak, O'Keeffe experienced a difficult climb in a snowstorm; this event was "recounted" in typical O'Keeffe-Price fashion in October 1880. Under the headline "Weather Bound: Wonderful Adventures of a Signal Service Officer on His Way to the Peak," the new tale began:

> Sergeant O'Keefe returned last evening from the unsuccessful attempt to ascend the Peak. He says it is the first time within his experience that he had been thus baffled, and he asks to be excused from ever being subjected to a similar experience. By the reports which he had received from Mr. Sweeney, who is stationed at the Peak, he was led to infer that but little snow had fallen and in consequence he was not prepared to contend with the obstacles which blocked his path.
> The journey for the first few miles of the trail was without any dissatisfactory feature, but while making an abrupt turn in the

trail in the vicinity of Minnehaha falls, the sergeant was brought to a standstill by an immense herd of black-tailed deer which impeded his further progress. He contends that the herd contained fully seven hundred head and says that it took just one hour and forty minutes for them to pass a given point.

The sergeant was armed only with a .32 caliber Smith and Wesson revolver and with this poor apology for a firearm he succeeded in killing seventeen of the deer. The only plausible reason that Sergeant O'Keefe can give for the remarkable appearance of this vast herd is that they were driven from South Park by the recent storm. As the sergeant was compelled to continue his journey to the Peak, he gathered the game which he had slaughtered, tied their tails together and slung them over the neck of his faithful mule, Balaam, and continued on his way.

Everything passed off smoothly until timber line was reached, when the sergeant encountered another serious barrier as the fierce northwest wind which accompanied the storm had formed mountains of snow across the trail. With the much trusted veteran mule "Balaam" and an unusual amount of perseverance on the part of himself, Sergeant O'Keefe contrived to surmount a tremendous snow drift 28 feet in depth. When safely upon the other side he paused for a moment, and taking his field glass he viewed the prospect o'er and examined the difficulties with which he had to contend. As far as the eye could reach nothing but snow banks could be seen, some of which were at least 100 feet in height. It required only a brief space of time for the sergeant to make up his mind that it would be useless for him to risk his life in making another rash attempt, so he concluded to return to the Springs, but upon turning to step into the saddle he discovered the mule had disappeared. The sergeant was now in a sad plight. Had he only survived the terrible rat raid of 1876 to find death again staring him in the face of starvation? He retraced his steps through that mammoth snow drift, and after a terrible siege of over one hour he found himself standing upon the other side utterly exhausted. After he had somewhat revived he glanced around him in hopes of finding some trace of the lost mule. And what was his chagrin to perceive the much trusted "Balaam" lying upon his back with feet uplifted in the air at the bottom of a deep ravine. The deer, with which he had been festooned, were scattered from top to bottom of the ravine. The sergeant secured the game and the mule and again started on his homeward journey.

While passing along a very secluded portion of the trail he was attacked by six ravenous mountain lions, and in order to save his own life he was compelled to cast away the game which had re-

quired so much exertion to capture. Even the seventeen deer did
not replenish their ravenous appetites and still they pursued him,
but by the proper manipulation of that mule O'Keefe managed to
evade them. He reached the signal office in this city at 8 o'clock
last night and it is doubtful whether he ever again will attempt the
Pikes Peak trail.[12]

Most of Colorado Springs enjoyed the tale; the number of doubters
and spoilsports was growing, however.

10.

A volcano "erupts." Private Jones is not amused. O'Keeffe loses a friend.
A fort is renamed. A new chief signal officer.
The study room.

DURING SERGEANT O'KEEFFE'S second tour of duty on Pikes Peak,
the daily journals, under his supervision now, did not exhibit the
literary excesses for which he had become famous. He had other
ways—cooperative reporters and newspaper editors—to present his
fantasies. On November 20, 1880, the editors of the *Colorado Springs
Gazette* introduced a story headlined "Fire, Smoke, Ashes and Melted
Lava Pouring from Crater" that gave further evidence the judge and
the sergeant had not lost their touch:

> The first knowledge that was given us of this peculiar and
> newly discovered phenomenon was reported last Saturday after-
> noon, and since that time a *Gazette* reporter [Eliphalet Price him-
> self?] under the guidance of Sergeant O'Keefe, visited the scene of
> what proved to be one of the most wondrous discoveries ever
> brought to light in the mountain region.
>
> Nearly all of the citizens of Colorado Springs have seen or read
> of the Crater which is located near the summit and just west of the
> Peak. It has always been conceded by scientists that this selfsame
> Crater had in times gone by been the scene of a terrible eruption as
> particles of lava had been discovered in the crevices of the rock ad-
> joining it.
>
> Those who have investigated more closely the various forma-
> tions which are peculiar to a volcanic mountain, generally have af-
> firmed that there are plenty of evidences to show plainly the Pike's
> Peak Crater has in its center a circular or cup-like opening through
> which lava has certainly been emitted.
>
> It was on the night of the 29th of October that the Crater first
> displayed any signs of volcanic activity. Sergeant O'Keefe was

aroused from his slumbers by a dreary, doleful sound which apparently emanated from beneath the signal station. His first convictions were that it was an earthquake, but this impression was soon dispelled by the fact that the sound still continued without any signs of a jar.

The sergeant concluded to investigate the cause of this mysterious sound and he, in company with his assistant, Mr. F. L. Jones, dressed themselves and started out in search of the cause thereof. They had barely stepped over the doorsill when a bright flash, at first thought to be lightning, surmounted the summit of the Peak.

It was only a second's duration and the Peak was again clothed in darkness. From this time on the sound heretofore described seemed to decrease until the usual quiet of the solemn mountain peak again was restored.

The following day Sergeant O'Keefe visited the Crater, feeling confidence that the sound heard on the night previous had emanated from that source. What was his surprise on looking down in the Crater to discover vapor curling up from the cup-like enclosure. This discovery only prompted him to further continue his researches, and after two hours laborious climbing he found himself standing within about 200 yards of the Crater chimney.

The heat even at this distance was very oppressive and the ground about him was covered with pulverized ashes and lava which had been emitted from what he believed to be an incipient volcano.

O'Keefe was lost in astonishment. The snow for a distance of half a mile from the Crater had entirely disappeared. This was all the more remarkable as it had upon the previous day been several feet in depth.

The sergeant was very much astonished at the remarkable discovery, thus brought to light, but he was not deceived by the calm. He was convinced that the absolute repose which the volcano then displayed would be of short duration. Since the 29th of October but one eruption has occurred and that was on the night of November 7th, when another one similar to that which occurred on the 29th, only more violent, occurred.

Sergeant O'Keefe happened to be up on the roof of the signal station on this occasion and he portrays the majesty of the scene as the grandest that he has ever witnessed, not excepting that of Vesuvius, seen by him in 1822 when he was a lad and before he left his native Italy for America.

The eruption began with a tremendous burst, which shook Pike's Peak to its very foundation, hurling into the air dense clouds of ashes and lava. These explosions succeeded each other with rapidity and increased violence for about an hour, when the volcano seemed to enter into a profound sleep.

During the eruption the clouds were strongly illuminated by the reflections of the glowing lava in the Crater, giving the scene the appearance of a vast conflagration. This will account for the peculiar light which has been noticed by the sheep herders on the plains east of the city.

Sergeant O'Keefe informs us that the flow of the lava tends towards Ruxton's creek, whence the water for the supply of the city is procured, and there is no doubt that the hot lava will, if it reaches the creek, so heat the water that it will be of no earthly use for drinking purposes.

It is evident that the eruption has but just begun, and should it continue any length of time there is no doubt that Colorado Springs will meet the same fate as that which destroyed the flourishing cities of Pompeii and Herculaneum.

The flow of lava has already extended a distance of three miles from the mouth of the crater and only two eruptions have taken place.

Scientists give it as their opinion that the present upheaval will last about three months, after which the volcano will settle down to a state of comparative repose, only to burst with renewed vigor in about six years.

The reporter who was sent to investigate this portentous development has not yet returned and fears are entertained for his safety.

The times were changing. The *Gazette* printed a partial disclaimer next to the volcano story. The editor at the *Denver News* stated he was certain that the good sergeant was born in Ireland and not in Italy in 1822; the *Denver Republican* pooh-poohed the entire account as an outright fabrication.

Poor Private Jones. Although the volcano story was just heating up, he was forced to telegraph Washington with the news that "Tuesday was one of the coldest ever experienced on the Peak. The thermometer showed a minimum of 36 degrees below zero during the entire night."[1] Fee Jones was uncomfortable in the company of Sergeant O'Keeffe. He overheard his commanding officer tell summit visitors the most outrageous stories. The two men were too different to get on. The private thought O'Keeffe's behavior unseemly and the sergeant thought Jones a bit of a prig. Jones let his friends in Colorado Springs and his colleagues in Washington know his low opinion of the Pikes Peak storyteller.

CHIEF SIGNAL OFFICER William B. Hazen had a turbulent six-year duty, 1880–1887. (*National Archives.*)

Sergeant O'Keeffe received a blow on December 17, 1880, when he learned that his old friend Eliphalet Price had died. Said the *Mountaineer* that afternoon: "It becomes our sad duty to chronicle the death of Judge Eliphalet Price, in this city, this morning. . . . He had been suffering for years with asthma, and the physicians ascribe his death to heart disease brought on by other trouble." The *Gazette* added a note to its obituary the following day: "He was a man of rugged character and a natural pioneer. He had a keen appreciation of humor and wielded a ready pen."[2]

Another death. Chief Signal Officer Myer died in 1880 at the age of fifty-two, shortly after he was made permanent brigadier general. Some memorialized Myer as the "Army's eyes, ears, and tongue."[3] Fort Whipple, Virginia, the first army installation to train Weather Bureau personnel, was renamed Fort Myer to honor the general.

His successor, General William Babcock Hazen, was, apparently, a kindred spirit to Myer. Hazen liked new ideas and the planning of ambitious ventures. He hired more civilian meteorologists to do research. At the behest of Cleveland Abbe, he instituted the "study room," whose purpose was to further "higher scientific work necessary to issue advancement in the art of weather prediction."[4] He wrote fresh instructions to volunteer observers stationed across the nation in order to convince them of their value to national weather forecasting. He immediately threw himself into projects that Myer had initiated and prepared for two ambitious expeditions to the Arctic, one to Point Barrow and the other to Lady Franklin Bay. These trips to the Northwest Territories of Canada in 1881 were called the First Polar Year, and were designed to initiate enough research on cold fronts so that the strength of their movements into southern Canada and the United States could be predicted. The tragic fate of one of these expeditions and a Signal Corps with sticky fingers around federal money would eventually thwart the ambitions of Chief Signal Officer Hazen, both for himself and for the Signal Corps's weather bureau.

11.

A LTHOUGH ELIPHALET PRICE WAS gone, John O'Keeffe continued to write tall tales, some of which were published. Often, the stories concerned the station's faithful but elderly mule Balaam and an unusual dog named Seldom Fed. The whitish appearance of the sides of the Pikes Peak summit inspired O'Keeffe to develop a story concerning the whitewashing of the peak. Historian Levette Davidson says that the sergeant was apparently very proud of this particular creation but, unfortunately, no copy has survived. Even though Colorado Springs was now a "sophisticated" town with 4,000 residents, most citizens continued to enjoy O'Keeffe's literary efforts. When supplies failed to be delivered to the weather station, the sergeant sent messages down the wire, one of which at least was picked up by the *Colorado Springs Gazette* on May 8, 1881: "Sergeant O'Keefe sustains his reputation for truth and veracity by claiming that he smoked ten pounds of gunpowder tea during his last sojourn on the Peak, the usual monthly supply of tobacco having failed to put in its appearance from Washington."

Stories of smoking volcanos and swarming rats seemed insignificant, however, compared to what O'Keeffe really saw on February 14, 1881. It had been foggy all day, and by evening

> four mock moons appeared on different sides of the moon . . .
> at equal distances from that body and from one another. These increased in intensity of colors until they were brighter than the moon itself, which appeared as if hidden by a veil.
>
> By and by, bows as brilliant as those of the sun appeared over each, which increased in size until they almost joined, and formed

NO "PIKES PEAK FEVER" HERE, apparently, for group posing in the morning sun. (*Western History Department, Penrose Public Library, Colorado Springs.*)

a perfect circle of the most brilliant colors. These gave way for double halos, the second being nearer to the first than that was to the moon. Both were of a beautiful violet tint.

When the moon had risen about two thirds of the distance between the horizon and zenith, these rings disappeared and a new one appeared, much further removed from the body. . . . The new halo was much paler than those it had succeeded; it was the precursor of one of the most magnificent refractions of light that any human being ever witnessed.

Near the moon appeared two mock moons, shining like balls of fire, and in the horizon opposite, these were reproduced, but in a milder color. A halo of mild pink color passed through the plane of these four mock moons and intersected with the main halo. To complete the display, a bow appeared in the zenith, and this was as brilliant as the rainbow, and comprised all the colors of that bow of promise.[1]

Although the more experienced of the weather observers felt that they had seen what there was to see on Pikes Peak, nature continued to surprise and amaze even the most jaded signalman. On May 11 Sergeant

O'Keeffe reported that hurricane winds had increased to 112 miles per hour by noon, at which point the cups of the anemometer were wrenched out and blown away. Two hours later, with no scientific way to measure the strength of the winds, O'Keeffe estimated their velocity at about 150 miles per hour.

Another surprise. On a warm evening in mid-June 1881, during a routine observation, the men went outside to watch what looked like a "luminous star, well defined . . . with a very short tail," moving through the sky at twenty-seven degrees.[2] They stayed up all night and successive nights to study the "star" which, at first light, slowly elongated into what they decided was a huge comet. On June 25 they could see the comet clearly; its nucleus was sharp and from it jetted three plumes stretching toward the moon. The comet's reddish-colored tail measured, to the eye, about ten feet long and four feet wide. Elsewhere, those with spectrascopes calculated that the comet's tail stretched to twenty degrees of arc at the height of its passage by the planet. The atmosphere surrounding the comet was hazy, noted the sergeant, and a few cirrus clouds seemed to float above the gaseous body as it moved along toward the northeast. As the men continued to watch the comet on successive evenings, however, it became fainter and fainter and finally disappeared.*

In July the weather men were treated to a number of violent thunderstorms; electrical charges snapped in the lightning arrester with such force that the observers feared the station would be destroyed. After one of these fierce storms, O'Keeffe noted in the log that a sheepherder named Thomas Douglass was found dead under a tree; the "lightning had stripped him of every stitch of clothing and his boots."[3]

On November 5, 1881, O'Keeffe made the *Gazette* again when the editors reported that the sergeant was digging an artesian well at the summit and had reached a depth of 628 feet. No further reports appeared of this ambitious undertaking.

At the end of the year, Colorado Springs residents were shocked to learn that John O'Keeffe had apparently been discharged from the U.S. Signal Corps. Evidently, Chief Signal Officer Hazen had received one too many complaints. True or not, one report to the papers explained that "due to what was described as a state of constant delirium tremens,

---

*Perhaps the signalmen saw the Tebbutt's Comet, which was sighted that summer by a number of astronomers.

THE FAMOUS MULE BALAAM jumps O'Keeffe's "lava flow." (*Western History Department, Denver Public Library.*)

O'Keefe received a dishonorable discharge from the Signal Service 'without character.' "[4]

Other newspaper articles of the day were much kinder to Sergeant John O'Keeffe. They gave details of a splendid banquet held to honor the noted storyteller; one celebrant toasted the sergeant, "his goblet filled to the brim with Iron Ute water. . . . O'Keefe, one of the greatest prevaricators, equalled by few, excelled by none. True to his record may his life be a romance and in his final resting place may he lie easy." Lieutenant H. P. Scott also proposed a toast: "The rosy realms of romance are as real to O'Keefe as the stern and sterile steppes of truth are to many. The golden glow which gilds the granite summit of the peak is but the type of that glamour which surrounds it through the mendacious genius of O'Keefe. Gentlemen, here's looking at you."[5] Hearty applause followed.

John O'Keeffe remained in the Colorado Springs area, at least part of the time. He worked for the Colorado Telegraph Company for a while and was also employed as a railway mail agent—doubtless, less rigorous jobs than that of weather observer at the top of a mountain. Even so, he died in 1895 at only thirty-nine, leaving a wife and two children. The gravestone of O'Keeffe's "baby daughter" Erin remained at the summit for a time, surrounded by the numerous boulders that geologist Frederick Hayden and his men had collected there in 1872. Sometime during the 1930s, however, some humorless executive ordered the gravestone removed. The venerable mule Balaam received better treatment; artist Charles Craig completed an oil painting of the celebrated animal, then in his late thirties, which was exhibited in the Colorado Springs Pioneers Museum.

If Private Fee Jones had cause for celebration because of the departure of O'Keeffe, he soon found he was not entirely free from eccentric sergeants. However much he suffered both shock and embarrassment from the outlandish behavior and lies of his former boss, his next sergeant, L. M. Dey, possessed a bit of whimsy himself. Dey reported that he and Private Jones were forced to leave a packer named Smith halfway along the trail to Pikes Peak during a fierce snowstorm. When they were able to come back for Smith one week later, Dey said the packer presented him with a bill to replace the supplies his two government-issued mules had eaten: "One wagon box, $3; one saddle, $25; two saddle blankets, $5; a portion of the stable, $5."[6]

As with earlier observers, Sergeant Dey had an opportunity to experience the wonders of St. Elmo's fire. At 8:45 P.M. on a June evening, toward the end of a heavy sleet and snowstorm, the telegraph wires were covered with bright, dancing lights.

> They presented the appearance of little electrified brushes of inverted cones . . . funnels of light with their points to the line, from which they issued in little streams about the size of a pencil lead, and of the brightest violet color, while the cone of rays was of a brilliant rose-white. . . ."[7]

When Dey ventured outside, he found that his hands lit up with flames of light. As he raised them, each finger was extended with a three-inch flame. He put one finger to his lips and heard a crackling sound but felt no pain. The slightly damp wristband of his woolen shirt was decorated with a ring of fire, as was his mustache. In ten minutes or so, the storm passed and the show was over.

T. HINE, PHOTO.

U. S. SIGNAL STATION, PIKES PEAK,

MANITOU SPRINGS.   ·   ·   ·   ·   ·   ·   COLORADO.

THE SELDOMRIDGE FAMILY made the climb to the summit in 1886. (*Photo by T. Hine;* *Starsmore Center for Local History, Pioneers Museum, Colorado Springs.*)

Two nights later, Dey heard singing sounds outside from the telegraph and the anemometer; he went out to investigate. The sergeant described the noise as that "produced by a carriage-wheel on a road-bed of hard snow on a cold frosty morning."[8] As he moved closer to the instruments, a bright violet light issued from each of the anemometer cups. Dey climbed to the roof of the station to better see the display. This was a different phenomenon, however, not like the harmless St. Elmo's fire of the earlier evening. It hurt his fingers. His hair stood up and prickled painfully. Dey's nose picked up the characteristic odor of ozone, the strong and penetrating smell that previous observers had noted with electrical events of this kind.

The sergeant climbed down from the roof, went to fetch his black felt hat for what he thought would be effective protection, and went up the ladder again to the top of the station house. This time he was "fairly raised off of his feet by the electrical fluid piercing through the top of his hat, giving him such a sudden and fiery thrust that he nearly fell off the roof in the excitement." Dey quickly snatched off his hat to see a beam of light run through it from top to bottom. "When the fluid began to thrust its fiery tongues into other parts of his body, he was spurred to

U. S. SIGNAL STATION, PIKES PEAK,
MANITOU SPRINGS.   .   .   .   .   .   .   COLORADO.

THE SELDOMRIDGE FAMILY back on burro and horse, ready for the trip down; note the ladies are riding sidesaddle. *(Photo by T. Hines, 1886; Starsmore Center for Local History, Pioneers Museum, Colorado Springs.)*

a hasty but 'brilliant' retreat.'"[9] The electrical display lasted about fifteen minutes but Sergeant Dey did not feel at all well for several hours following his visit to the roof.

It finally happened. After nine years of occasional lightning strikes, which usually knocked out the arrester, destroying the wires leading into the station, filling the rooms with smoke and terrifying both observer and visitor, the weather post suffered a direct hit at 4:31 P.M. on July 1, 1882, during a heavy hailstorm. The explosion was deafening. Three men were inside the building; Private W. R. Boynton and civilian worker E. K. Cribbin were badly burned and bruised, and Sergeant Dey was numb from having been thrown about. The log for that day made vivid reading:

> The fluid passed through the outer partition walls and entered the office in the southeast corner near the stove, tearing up the floor, melting and tearing off the zinc sheeting around the stove, jumped to the self-register, which it demolished, also the regulator clock on the wall, burned up completely the office wires, and, passed out at the north window to the roof, burned out the dial of the anemometer.[10]

All the windows of the weather station were broken. Later, the stunned men found out that, at Manitou Springs below, a cloudburst had knocked out a number of bridges and flooded almost every house in town. After the lightning strike, Mr. Cribbin's health deteriorated to such an extent that Sergeant Dey notified Washington that he feared the observer's injuries would be mortal. Cribbin had been carried down to Colorado Springs by stretcher. He weakened day by day and on August 13 he died. Dey collected Cribbin's personal effects to send to his family. Not only was Cribbin replaced by a new man but, two months later, Sergeant Dey left the peak as well.

For some years, there had been talk in Washington of building a new station at Pikes Peak. When Lieutenant McClellan inspected the building in 1879, he recommended to General Myer that this be done "and a more cheerful and comfortable building put here on this bleak spot, where the men employed are exposed to so much privation and hardship." At the very least, "It is necessary and proper that a Privy should be constructed here for the use of men on station."[11] After the 1882 lightning strike, Lieutenant W. A. Glassford was dispatched to the Colorado Springs area to evaluate the station and its future value to the Signal Service. Glassford, already famous for his work with military balloons, decided that it would be prudent to start from scratch with the laying of telegraph wire to the summit instead of the previous mend-and-patch method, and that it would be equally prudent to start over with a new station, instead of remodeling the old one.

The lieutenant was not in the mood to dally and quickly awarded a construction contract for a new building to architect Lewis F. Dey of Philadelphia, who happened to be visiting his brother, Sergeant L. M. Dey, at the Springs. The work was started on August 7, a little over a month after the disastrous lightning strike. New stone walls were built, then mortared, and then torn down. Evidently, the masonry work was inferior and Lieutenant Glassford ordered that it be replaced. He decided to stay in the area until the work was completed to his satisfaction. On October 13 the men moved their belongings and weather equipment into the new structure. The rectangular building, which measured forty-four by twenty-nine feet, included an office, a kitchen, a dining room, two bedrooms, and an outside privy. Elsewhere, new telegraph poles were set and better-quality wire was ordered and installed. Surveying a working telegraph and a new and more secure building, satisfied that his work was finished, Lieutenant Glassford returned to Kansas and his military balloons. Alas, on October 19, the new telegraph line went down. From then on, Washington was lukewarm on

WINTER AT THE NEW weather station, sometime after 1882. North wall shows new instrument shelter; left roof houses anemometer; at center is wind vane and rain gauge; snow drifts on west wall. (*Western History Department, Penrose Public Library, Colorado Springs.*)

the subject of line repair at Pikes Peak; the telegraph was fixed only occasionally after 1882. The new station itself proved to be less than perfect as well. The roof leaked, the mortared walls were still not secure enough to prevent snow from entering the building; nor were the newly fitted windows and doors. Even so, the men knew their living and working quarters had improved from the old building; after all, they did have a privy now. The following August, the old station was torn down.

After a busy 1882, Sergeant Harry Hall was posted as chief observer and apparently enjoyed a long and untroubled stint of four years at the peak. Sergeant Seyboth would have loved to carry out the orders given Hall in 1883. He was to hire a cook, lay in food supplies, and feed those visitors to the summit who arrived at mealtime. They were to be charged, of course, said Washington. The new sergeant tried unsuccessfully to snowshoe to the peak but, unlike John O'Keeffe, he did not try to make a story of it. During Hall's tenure, the tin roof of the mule stable blew off during a windstorm; part of the roof of the main building was damaged as well. Both were repaired but Hall complained to Washington that he believed the winds had damaged his anemometer so that it was not recording accurately.

FOUR HIKERS REST ON the north side of the station before the trip down. (*Western History Department, Denver Public Library.*)

In late May 1883, Hall studied a number of rings that formed around the edge of the sun and were evident for most of the summer. The rings were colored reddish brown, he said, and did not seem to fit any of the categories of coronas or halos outlined in the Signal Corps manual. Sergeant Hall was puzzled. Elsewhere on the planet—on a remote island in the Sunda Straits lying between Sumatra and Java, twelve thousand miles from Pikes Peak—an active volcano called Krakatoa had erupted on May 20. It was quiet for a time and then erupted again on June 19. During August 26 and 27, Krakatoa had its grand finale, sending an immense amount of black ash seventeen miles into the atmosphere and darkening the skies as far away as Madagascar. The resulting tidal wave along nearby coasts caused much destruction and loss of life. The volcanic ash from Krakatoa drifted around the earth a number of times that summer, causing brilliant bloodred sunrises, equally spectacular sunsets, and greenish suns and moons. What Sergeant Hall had seen from the top of Pikes Peak were Bishop's rings, which edge the sun's corona after violent volcanic eruptions, a phenomenon seldom observed.*

---

*The rings were named after the Reverend Sereno E. Bishop of Honolulu, who first described such an event.

AFTERNOON VISITORS TO THE SUMMIT. Note building extension at south end. *(Photo, Western History Department, Penrose Public Library, Colorado Springs, Colorado).*

General Hazen's term as chief signal officer was but six years, a short stint compared to General Myer's twenty years. However, those six years were filled with Senate investigations, charges and countercharges in the *New York Times,* an accusation of racism, mistreatment of Signal Corps men, and allegations of personal fraud linked to General Hazen himself. Few alleged crimes were left out of the complaints, apparently.

Hazen's problems probably went back as far as the Civil War when he was court-martialed for alleged cowardice in action. During his tenure as chief signal officer, which started so positively, the polar expedition to Lady Franklin Bay in 1881 met with the worst sort of disaster. Lieutenant A. W. Greely and a party of his men were stranded when the relief ship, *Proteus,* wrecked. A number of the men perished. No provisions were available for those who were left at Lady Franklin Bay; in fact, the group whose responsibility it was to rescue the Greely party left the area. Eventually, in 1884, the group left behind made its way back to civilization. Such events made excellent fodder for the newspapers and for those who were against the weather bureau's continued military administration. A black graduate of the City College of New York with apparent impeccable credentials was denied employment by the general, an action that irritated a number of his military colleagues.

Secretary of War Robert Todd Lincoln had never been a fan of the Signal Corps. He didn't like Myer before and he didn't like Hazen now. He particularly didn't like the fact that the Signal Corps budget was taken out of his congressional allotment of funds. Said Lincoln: "The Signal Corps has a professor in charge of meteorology, a professor in charge of the electrical department, and various junior professors and clerical employees. I think these people could take charge of that business, and it should be a civilian force. . . ."[12] In other words, get them out of the War Department. One wonders what words Secretary Lincoln used to describe Cleveland Abbe's "study room."

In the halls of Congress and in the newspapers, stories circulated about the mistreatment of Signal Corps men—poor training, poor food, low salaries, and little chance for advancement. Record keeping was shoddy; audits were faked, and so were some of the weather records. Calls for reorganization were heard, but nothing much was done about it. Hazen's real problem, however, was not of his making. Captain Henry W. Howgate, the man to whom Robert Seyboth had sent a beautiful bird and marmot years before, had been systematically embezzling funds from the Signal Corps with the help of phony vouchers. In fact, he had apparently started those activities long before General Hazen took command. Howgate was arrested in 1881. At first, it was announced to the papers that Howgate had stolen some $40,000. Then it was $90,000. The final reckoning, however, proved to be $237,000. Howgate was jailed; however, he broke out, and was at large for some years. It was no surprise, then, that Congress cut the budget of the Signal Corps in 1885, a move that forced the closure of eighteen weather stations. Many continued to call for reorganization of the weather bureau.

The little Pikes Peak station remained open, but not for long. Naturally, the situation in Washington affected the morale of the signalmen throughout the country. Their leaders became more timid and more defensive than usual, while the budget for supplies and the replacement of instruments was stripped to the bone.

When General Hazen died in 1887, the lieutenant who had been left behind on an icy shore of northern Canada was now a general. Adolphus Washington Greely took command of the Signal Corps. The new chief signal officer had published a lengthy book called *American Weather* in which he wrote about man's eternal fascination with weather phenomena. He noted in the introduction, "From the beginning of time the alternation of the seasons and the irregular recurrence of weather conditions must have interested man and engaged his attention. The very

A FREQUENT VISITOR to the Pikes Peak
station, Lieutenant Adolphus W. Greely
led an ill-fated expedition to Lady
Franklin Bay, Northwest Territories,
1881. (*National Archives.*)

AN OLDER A. W. GREELY, now Brigadier
General and Chief Signal Officer of the
Signal Corps (1887–1891). (*U.S. Army.*)

ancient Book of Job and the later books of the New Testament contained formulated weather wisdom, the result of man's primitive observation."[13] It was General Greely's goal to take what he regarded as an
equally primitive weather observation team and transform it into a
modern sophisticated weather bureau. He would transform it out of existence, at least as far as the Signal Corps was concerned. And the station at Pikes Peak? Chief Signal Officer Greely blew hot and cold on
pure meteorological research and took the surprisingly shortsighted
view that the study of atmospheric electricity and its relation to weather
forecasting was not of sufficient value to maintain a high-altitude observer station under his command.

HALFWAY HOUSE WAS greatly expanded by 1888. *(Starsmore Center for Local History, Pioneers Museum, Colorado Springs.)*

12.

A flight of moths. Timothy Sherwood tries to hold the line.
A carriage road is built. A homestead with a lunch counter. Gold fever.
A railroad in the sky. A hymn is written. Unusual ways to the peak.
AdAmAn sets off fireworks. The old is torn down. On with the new.
Contributions of the weather station.

T HE ARRIVAL OF LITTLE brown and gray moths on May 14, 1888, un-
derlined the feeling of foreboding those connected with the station
were experiencing. Corporal Timothy W. Sherwood noted in the log
that these small creatures clouded the sky, "flying with the wind from
north to south over the peak and valley to the east of the summit," a not
unusual occurrence along the Front Range.[1]

The following day, Sherwood noted the passing of the season's first
thunderstorm accompanied by the noises that by now seemed common-
place to the men at the summit. The corporal reported the presence of
that "peculiar buzzing sound . . . coming from every exposed point
of the building and from the rocks, wet by the melting hail."[2]

When Lieutenant John C. Walshe inspected the station in August
1887, he was distressed to find untidy rooms, including the kitchen;
moreover, he reported to his superiors that the men had not even made
their beds. He did allow that Sherwood kept pretty good records, however, particularly of snowfall data. Later that year, despite rumors that
Washington no longer found the Pikes Peak station useful, the facility
was renovated and refurbished. Walls were reinforced, windows were
lined to keep out the cold and sleet, and the shabby and outmoded scientific equipment was brought up to standard. Once the roof was repaired
again, Washington ordered Sherwood to keep visitors from climbing to
the top of the station as the resulting wear and tear produced more leaks
and the need for more repairs.

Sherwood was the last signalman to supervise the weather station.

NEW INSTRUMENTS FOR a dying station. Left to right: accumulating anemometer, three-cup anemometer, wind vane, rain gauge, and a refurbished instrument shelter on the west wall of the building. *(National Archives.)*

The telegraph was long gone. Nevertheless, the men continued to keep up the log with careful notes on changes in the weather as well as the mention of unusual plants and animals sighted. But Captain Howgate and his phony vouchers had caused considerable bad feeling in Congress; not only were the lawmakers loath to appropriate additional money for the Weather Bureau, but also they slashed the existing budget. A further number of observer posts was closed due to money limitations and the station on Pikes Peak was on the list. Corporal Sherwood was ordered to store all heavy observation equipment and the stove in one padlocked room in the building. He brought the lighter equipment down the trail to the Colorado Springs office. The mountain outpost had been dedicated with great fanfare fifteen years before and memorialized in an article in *Harper's Weekly*. No notables were on hand, apparently, on October 18, 1888, when the station officially closed.

Sherwood remained on duty, however, for a few more years. During his tenure, the signalman must have felt he was under seige from a number of quarters. Orders from Washington directed him to supervise the work at the peak, then to go down to the signal office at the Springs,

IT'S JUNE 1889 but the summit is still snowy. (*Western History Department, Penrose Public Library, Colorado Springs.*)

and then back to the peak. He was also frustrated by the waywardness of his assistant, Private C. F. Schneider, who sometimes appeared at whatever workplace to which he had been assigned, and sometimes not. The private obeyed some orders and, apparently, completely ignored others. Sherwood complained to Washington that he was asked to undertake additional work for the Colorado State Weather Service; furthermore, he was not being paid regularly.

In August 1887 Sherwood began what would be a lengthy and confusing correspondence with various Washington authorities on the subject of the building of a private wagon road that would encroach on the Pikes Peak Military Reservation. He alone was ordered to stop the road crew from entering army property while Washington bureaucrats argued back and forth as to the propriety of a private company building a wagon road on government land. Further, there appeared to be considerable confusion as to the exact boundaries of the Pikes Peak Military Reservation itself. On August 10, 1888, Sherwood fired off the following telegram:

> Notified wagon road Co. they must do no work on reservation to-day. They refuse to comply. One hundred and twenty men now at work on reservation. Sixty more tomorrow. I await further orders.[3]

AN 1893 VIEW OF THE toll road to the peak, built in 1888, despite Corporal T. W. Sherwood's objections. (*Photo by William Henry Jackson.*)

A variety of orders came Corporal Sherwood's way, most of them contradicting one another. Meanwhile, he continued to tell Washington that the wagon road was proceeding and that he alone was powerless to stop the large road-building operation that was moving its equipment up the trail under cover of fog. It would seem that Chief Signal Officer Greely had washed his hands of the affair and was waiting for instructions from either the president or Congress. On August 27 the corporal sent another letter to General Greely:

> . . . The wagon road is completed to the summit, carriages are now driving up to front door of the station. There seems to me to be but one way now left to punish the road company for an open defiance of the service by placing the guard referred to within. Travel over the road is very heavy and the company are deriving an income from the tolls collected. This action would necessitate the sending here of four armed men for the duty.[4]

Timothy Sherwood's pleas for clarification fell on deaf ears or, at least, on bureaucratic ears. Toward the last, nature joined the confusion with an August 20 lightning strike through the station. Two patrons of the new toll road, George I. McClure of Gibson City, Illinois, and Laura M. Cook of Chicago, were badly burned; the following day, after receiving emergency medical treatment, both victims were carried on cots back down the wagon road.

Evidently, neither General Greely nor the Department of Agriculture, which took over management of the Weather Bureau in July 1891, could take a firm stance with regard to the summit station. In January 1889 Greely directed that the weather building be sold at auction; later, he rescinded the order. On September 8, 1892 the mountain observer post opened again with ambitious plans for refurbishment, which included the construction of a second story and the rental of rooms to visitors at weekly rates not to exceed eight dollars. In time, the proposals went awry, and Timothy Sherwood turned over the building to the Department of the Interior and was assigned other duties. The station opened and closed again, at least on government paper; by 1896, however, the federal government had lost interest in the site, apparently.

In Washington, the weather bureau continued to expand and develop more sophisticated tools for the prediction of storms. Former army observers were given preferred status in the civilian agency or were allowed to leave the military. The bureau remained with the Department of Agriculture until June 30, 1940, when it moved again to the Department of Commerce.

Though the weather men had left, the peak continued to be used for a variety of purposes. For example, Alfred G. Lewis, a Manitou Springs physician and sometime mayor, filed a homestead claim on February 6, 1889. He built a rude log cabin a few feet from the old weather station to fulfill at least part of homesteading requirements. He then prepared to fulfill the agricultural requirements as well. The doctor and his friend Harry Dauchy laboriously hauled dirt up the trail to the peak and "spread it in sheltered, sunny areas among the rocks." The men planted corn, wheat, oats, and potatoes. "The corn and wheat grew a few inches high, but, of course, never matured. The potato plants leafed but no tubers developed."[5]

Dr. Lewis was not serious about farming on the summit of a 14,000-foot mountain. But he was serious about the business opportunity occasioned by the many tourists coming up the trail. The would-be homesteader wanted to be certain that his patent to the land met all the federal rules. He laid in supplies and served coffee and doughnuts to the tired climbers, expanding his service, after a few seasons, into a sort of lunch counter. Dr. Lewis continued in business there for five or six years.

Zalmon Gilbert Simmons was not at all happy about Dr. Lewis' lunch counter of his homestead. Simmons, who had made a great deal of money in Wisconsin cheeses and telegraph wire for Western Union, achieved later fame with the manufacture of "Beauty Rest" mattresses. The financier wanted to build a railway to the top of Pikes Peak. It had been tried once before but the venture went broke. A comfortable train to the top of the mountain was very appealing to Simmons, who had taken his first ride up the trail, dressed in formal frockcoat and silk hat, riding a mule—the famous Balaam, of course.

Money was no obstacle to Simmons, but a legal homestead of one hundred and sixty acres was. Negotiations began. Dr. Lewis felt that $25,000 was an appropriate sum for his summit rights. Simmons disagreed strongly and offered him $5,000, but the doctor immediately turned him down. Highly vexed, Simmons sent his attorney to Washington to investigate the claim. Ultimately, Dr. Lewis and his lunch counter lost. The homestead claim was ruled to be illegal, the Manitou Springs doctor never received the five thousand dollars, Simmons got his way and negotiated a ninety-nine-year lease with the federal government.

Simmons capitalized his venture as the Manitou and Pikes Peak Railway Company on November 14, 1888, just one day short of the anniversary of Zebulon Pike's sighting of a little blue cloud through his telescope eighty-two years before. The 8.9-mile cog "railroad in the sky,"[6] modeled after the Swiss Abt system, was completed two years later, but not without difficult solutions to tricky engineering problems. Simmons engaged the consulting services of Wilhelm Hildebrand, whose firm built the Mt. Riga cog railroad in Switzerland, as well as Washington A. Roebling, the engineer who had directed the construction of the Brooklyn Bridge. Together, they planned for a railroad bed that would have an elevation of 846 feet per mile, with an average grade of 16 percent. It took three locomotives to push the single passenger car to the summit.

The most serious problem in the construction of the railroad was a human one—the continual irritability and recurring illnesses among the crew. It seemed that Pikes Peak fever affected them just as it had the weather observers earlier. In addition, it upset the construction workers to have St. Elmo's fire dance in their hair and on their fingers while they attempted to work on the rail bed just after a storm. They had to insulate some of their tools with rubber hosing to avoid serious shocks as well.

Another kind of fever was evident in the Pikes Peak neighborhood in 1891. Gold was finally discovered in nearby Cripple Creek, the spot that

THE FIRST TRAIN TO Pikes Peak stopped at the eastern entrance, 1891. *(Photo by J. G. Hiestand; Western Historical Collections, University of Colorado at Boulder.)*

Professor Hayden had pinpointed in 1872 as a perfect site for mineral exploitation.

Teddy Roosevelt was one of the early riders on the "railroad in the sky;" after his trip, he pronounced that the scenery in the area "bankrupts the English language."[7] In 1893 another visitor to the summit, Katharine Lee Bates, was so awed by the grand sight that she composed a hymn:

> O beautiful for spacious skies,
> For amber waves of grain,
> For purple mountain majesties
> Above the fruited plain!
> America! America!
> God shed his grace on thee
> And crown thy good with brotherhood
> From sea to shining sea!

In 1896 one of the cog railway's locomotives got loose, careened down the mountain and crashed; fortunately, no one was hurt, but thereafter the engines were equipped with automatic brakes. In 1947 the railway converted to six diesel-electric trains; each could comfortably seat fifty-six persons. Clearing the tracks in the spring was, of course, a continual problem, solved with the judicious use of dynamite and a wedge snow shovel. Not until 1954 was a rotary snow plow put into op-

NO LONGER A WEATHER STATION, Summit House was host to those who came up by train. Note the elaborate observation platform, built in 1899. (*Western History Department, Denver Public Library.*)

eration. In 1963 new self-propelled trains of Swiss manufacture were installed; each train could carry eighty passengers. The cars went to the peak at five or six miles per hour—considerably faster than a person could ascend on foot or on the back of old Balaam.

Through the years, it became fashionable to make the trek up to Pikes Peak in some unusual way. In 1888 Mrs. John Wesley Powell, covered with a green veil and three petticoats, accompanied her geologist husband to the summit. On August 12, 1901, W. B. Felker and C. A. Yont barely made it to the top in a brand new Locomobile Steamer; since Mr. Yont spent most of the fifteen-hour trip pushing the little car over rough spots in the road, it is not difficult to understand why he collapsed half a mile from the summit. The Felker and Yont auto climb was followed by a successful attempt by a twenty-horsepower Buick Bearcat in 1913. Fitting in with the zany pranks of the 1920s, "Peanut Bill" Williams, from Rio Hondo, Texas, crawled to the top on March 25, 1929, pushing a peanut with his nose; it took him twenty-two days. Larry Hightower made the trip in just four days pushing a wheelbarrow in 1949. Soon, horse racing to the summit became a regular event, as were footraces. After a new highway was completed in 1916, stock car and motorcycle racing up Pikes Peak became yearly contests. A far cry from Private Smith's struggles through the snow in the 1870s.

A group of five hundred Masons held a private service on the summit in 1899 and installed a hundred-and-fifty-pound solid-brass plaque. Some years later, the memorial plaque was stolen and has never been recovered. One couple found the summit a perfect site for their July 14, 1904 wedding; luckily, the weather cooperated. At some point, an unknown visitor to the top of the mountain decided to end his life on the summit; he is buried there with no marker.

In 1922 five men decided they would hike up the mountain on the last day of the year, snow drifts or sleet storms notwithstanding. They made the summit handily but with frosty fingers, so they fired up the old weather station stove that had given Sergeant Seyboth a number of nasty shocks in earlier days. As they warmed themselves, they toasted the new year, their glasses filled with ice water. Finding some flares left over from the railway construction, the men lit them and sent the torches into the air, causing a few bad moments for residents below. The line to the fire department was busy that night with reports of strange goings-on at the peak. Each year since that time, the group, which adds one new member each trip—hence, they call themselves AdAmAn—makes the trek on New Year's Eve. They pack in halfway by horse, climb the last twelve miles before midnight, and set off elaborate fireworks. Some Colorado Springs residents return the compliment and flash lighted mirrors from below to let the men know they are watching.

For a number of years, science researcher Kathryn Shapley attempted to interest public and private groups in saving the old weather station as an historic monument to the weather pioneers of the Signal Corps. She interviewed prominent Coloradans, wrote many letters, and made numerous telephone calls, to no avail. The main deterrent to preserving the station was the lack of communication between various jurisdictions supervising activities at the peak—the city of Colorado Springs, the U.S. Forest Service, and nearby Colorado State University. In 1963 the "shabby old structure"[8]—the thick-walled stone house that Chief Signal Officer William B. Hazen had ordered built in 1882 to replace the earlier, even shabbier structure—came down. A modern visitors' center was then constructed at a cost of $500,000. The newer structure settles and cracks from thermafrost and must be repaired every few years.

Today, the high-altitude observatory at Mount Washington still flourishes. The National Center for Atmospheric Research, established in the 1950s in Boulder, Colorado, plays a vital role in weather science, as does the National Oceanic and Atmospheric Administration, a federally funded group also based in Boulder. The University of Colorado maintains high-altitude stations that include the Institute of Arctic and Alpine Research, which has operated the Mountain Research Station since 1951, although the station itself dates back to 1914. The technology of these institutions is awesome; their research is far-reaching and continually increases our knowledge of weather prediction. Their ability to promulgate their findings is far superior to the daily attempts to transmit bulletins by way of a broken-down telegraph line from the summit of a stormy peak. The layman, however, wishes for better weather prediction at times. Mother Nature continues to surprise us with flash floods, ice storms, violent winds, and blinding snows. Death, the destruction of crops, and sinking ships still occur when fierce weather arrives without notice.

At Pikes Peak, nothing is now left to give evidence to visitors that, before the turn of the last century, U.S. Army Signal Corps observers studied storms and other weather phenomena there, transmitting their findings to Washington. Even though they have not been accorded the credit, their reports eventually provided the basis for scientific advances in the study of meteorology. But not without cost. One signalman died after being struck by lightning. Another became incurably insane. Most suffered with the aftereffects from frostbite on their hands, ears, and feet; their eyes were damaged from repeated bouts of snow blindness. They lived removed from society for long periods. They put up with inadequate sanitation, poor food, low pay, and bureaucratic whims from Washington. Despite these difficulties, most of the men survived with some grace. Modern meteorologists know that early researchers like Benjamin Franklin, Thomas Jefferson, Joseph Henry, James Espy, Increase Lapham, and Cleveland Abbe were the pioneers in their field. Perhaps they should add a succession of eccentric but intellectually curious army enlisted men who manned the tiny stone station on Pikes Peak.

Hikers in the Pikes Peak area still cross the stones of a little stream called Boehmer's Creek, named for the sergeant who broke his finger while attempting one of the first pack trips to the summit. Nearby, two miles south of Pikes Peak, is 12,590-foot Sackett Mountain, named for the sergeant who was so excited to find a shrimp in Mystic Lake. The Pikes Peak weather observers of the 1870s and 1880s are nearly forgotten, but their legacy lies in the advanced research on high-altitude phenomena now being done in the sophisticated scientific laboratories and observatories located near the original mountain station.

NOTES

CHAPTER ONE

1. David Lavender, *The Rockies* (New York: Harper and Row, 1968), p. 241.
2. Ibid.
3. Joseph M. Hawes, "The Signal Corps and Its Weather Service, 1870–1890," *Military Affairs* 30 (Summer 1966), p. 68.
4. "Meteorological Observations Made on the Summit of Pike's Peak, Colorado, January 1874 to June 1888," *Annals of the Astronomical Observatory of Harvard College* 22: vii–vix, 459–75 (Cambridge, Mass.: John Wilson and Son University Press, 1889).

CHAPTER TWO

1. Edwin James, *An Account of an Expedition from Pittsburgh to the Rocky Mountains Performed in the Years 1819, 1820* 2: (London: Longman, Hurst, Rees, Orme, and Brown, 1823), p. 215.
2. Robert Athearn, *High Country Empire: The High Plains and Rockies* (Lincoln: University of Nebraska Press, 1960), p. 4.
3. W. B. Vickers, "History of Colorado," in *History of Clear Creek and Boulder Valleys, Colorado* (Chicago: O. L. Baskin and Company, 1880), p. 19.
4. Zebulon Montgomery Pike, *An Account of Expeditions to the Sources of the Mississippi, and through the Western Parts of Louisiana, to the Sources of the Arkansaw, Kansas, La Platte, and Pierre Juan, Rivers* (Philadelphia: C. and A. Conrad and Company, 1810), p. 169.
5. James Expedition, p. 215.
6. Ibid., p. 220.
7. Quoted in John Fetler, *The Pikes Peak People: The Story of America's Most Popular Mountain* (Caldwell, Idaho: Caxton Press, 1966), p. 54.
8. George F. Ruxton, *Adventures in Mexico and the Rocky Mountains* (New York: Harper and Brothers, 1860), pp. 246–47.
9. Julia A. Holmes, *A Bloomer Girl on Pikes Peak—1858,* ed. Agnes Wright Spring (Denver: Denver Public Library, 1949), p. 13; p. 16; ibid.; ibid.
10. Ibid., p. 34; p. 35.
11. Ralph Waldo Emerson, "Friendship," in *May-Day and Other Pieces* (Boston: Ticknor and Fields, 1867).

12. "History of Colorado Springs," undated paper, no author listed, p. 1.

13. Henry Villard, *Past and Present of the Pikes Peak Region, 1860* (Princeton, N. J.: Princeton University Press, 1932 reprint), p. 105.

CHAPTER THREE

1. George S. Bliss, "The Weather Business: A History of Weather Records, and the Works of the U.S. Weather Bureau," *Scientific American* 84, supplement no. 2172 (August 18, 1917), p. 110.

2. Philip D. Thompson and Robert O'Brien, *Weather* (New York: Time, 1965), p. 157.

3. Eric R. Miller, "New Light on the Beginnings of the Weather Bureau from the Papers of Increase A. Lapham," *Monthly Weather Review* 59 (1931), pp. 65–70.

4. U.S., Congress, House, *Letters of Hon. Halbert E. Paine,* Exec. Doc. no. 10, pt. 2, 41st Cong., 2d sess., 1870, pp. 2–3.

5. *Army Times,* eds., *A History of the U.S. Signal Corps* (New York: G. P. Putnam's Sons, 1961), p. 56; ibid.

6. *Letters of Paine,* p. 4.

7. *Army Times, History of Signal Corps,* p. 57.

8. Ibid., p. 59.

9. U.S., War Department, "Practical Use of Meteorological Representations and Weather Maps" (Washington, D.C.: Government Printing Office, 1871), p. 37.

10. Fetler, *Pikes Peak People,* p. 133.

11. Harry Hansen, ed., *Colorado: A Guide to the Highest State* (New York: Hastings House, 1941), p. 111.

12. Ibid., p. 114.

13. Isabella L. Bird, *A Lady's Life in the Rocky Mountains* (Norman: University of Oklahoma Press, 1960), p. 152.

CHAPTER FOUR

1. James H. Smith, *Our Country: West* (Boston: Perry Mason Company, 1912), p. 124.

2. Ibid., pp. 124–25.

3. Fetler, *Pikes Peak People,* p. 156.

4. Smith, *Our Country,* p. 125.

5. Ibid., p. 126.

6. Fetler, *Pikes Peak People,* pp. 162–63.

7. John Gibson, *The Trail 1,* no. 10, 1917.

CHAPTER FIVE

1. *Harper's Weekly,* November 8, 1873, p. 989.
2. Ibid.
3. U.S., Signal Corps, *Annual Report of the Chief Signal Officer to the Secretary of War, 1874* (Washington, D.C.: Government Printing Office, 1874), p. 57.
4. U.S., Signal Corps, *Report of the Chief Signal Officer to the Secretary of War,* paper no. 6, October 1873 (Washington, D.C.: Government Printing Office, 1873), p. 114 (hereafter cited as *Paper no. 6*).
5. Ibid., p. 118.
6. Smith, *Our Country,* p. 127.
7. Ibid.
8. *Paper no. 6,* p. 118.
9. Smith, *Our Country,* p. 127.
10. *Paper no. 6,* p. 114.
11. Ibid.
12. Ibid.
13. Ibid.
14. Ibid.
15. Ibid., p. 120.
16. Ibid., p. 115.
17. Ibid.
18. Ibid.
19. Marshall Sprague, *Newport in the Rockies* (Denver: Sage Books, 1961), p. 69.

CHAPTER SIX

1. *Paper no. 6,* p. 116.
2. Ibid.
3. Smith, *Our Country,* p. 128.
4. *Paper no. 6,* p. 116.
5. Ibid.; ibid., p. 117.
6. Ibid., p. 118.
7. "Meteorological Observations," p. 461.
8. *Paper no. 6,* p. 115; "Meteorological Observations," p. 459; ibid.
9. *Paper no. 6,* p. 117.
10. "Meteorological Observations," p. xi.
11. U.S., Signal Corps, *Report of the Chief Signal Officer to the Secretary of War, 1882,* appendix 72, "Extracts Relative to Electricity, from Pikes Peak Monthly Journal" (Washington, D.C.: Government Printing Office, 1882), p. 894.
12. Ibid, p. xi.

13. *Paper no. 6*, p. 119.
14. Ibid.
15. Ibid., p. 120.
16. Ibid.
17. U.S., Department of Commerce, Weather Bureau, *Weather Glossary*, compiled by Alfred H. Thiessen (Washington, D.C.: Government Printing Office, 1946), p. 228 and Great Britain, Meteorological Office, *Meteorological Glossary*, compiled by D. H. McIntosh (London: H.M. Stationery Office, 1963), p. 219.
18. *Paper no. 6*, p. 121.
19. Ibid.; ibid., pp. 121–22.

CHAPTER SEVEN

1. *Paper no. 6*, p. 124.
2. Phyllis Smith, *History of Floods and Flood Control in Boulder, Colorado* (Boulder: City, September 1987), pp. 13, 52a, 55, 66, 74.
3. Robert O. Rupp, *Pikes Peak Duty: A Brief History of the Pikes Peak Military Reservation* (Fort Collins, Colo.: R. O. Rupp, 1987), pp. 42–43.
4. *Paper no. 6*, p. 124.
5. Ibid., p. 122; 122–23.
6. Ibid., p. 124.
7. Ibid.
8. Ibid., p. 123.
9. Ibid., p. 122; ibid.
10. Ibid., p. 124.
11. Ibid., p. 125.
12. Ibid.; *Colorado Springs Weekly Gazette*, June 27, 1874.
13. *Annual Report, 1874*, p. 58.
14. Ibid., p. 57.

CHAPTER EIGHT

1. U.S., Signal Corps, *Report of the Chief Signal Officer to the Officer of War, 1875* (Washington, D.C.; Government Printing Office, 1875), p. 60.
2. "Meteorological Observations," p. 461.
3. Fetler, *Pikes Peak People*, p. 174.
4. Ibid., p. 183.
5. John Timothy O'Keeffe, "Attacked by Rats, Terrible Conflict on the Summit of Pikes Peak," *Pueblo Chieftan*, May 25, 1876.
6. Fetler, *Pikes Peak People*, p. 183.
7. Levette Jay Davidson, "The Pikes Peak Prevaricator," *Colorado Magazine* 20 (November 1943): 218.

8. Felter, *Pikes Peak People*, p. 181.
9. Ibid., p. 82.

CHAPTER NINE

1. "Meteorological Observations," p. 462.
2. Ibid., p. 464.
3. U.S., Department of Commerce, Weather Service, daily log of the Pikes Peak Signal Station, 7 February 1877, Record Group 27, National Archives, Washington, D.C.
4. Great Britain, *Meteorological Glossary*, p. 288.
5. "Meteorological Observations," p. 462.
6. Ibid., p. 468.
7. Ibid.
8. Ibid., p. xii.
9. *Army Times, History of Signal Corps*, p. 61–62.
10. U.S., Signal Corps, *Annual Report of the Chief Signal Officer to the Secretary of War, 1878* (Washington, D.C.: Government Printing Office, 1878), p. 68.
11. U.S., Signal Corps, *Annual Report of the Chief Signal Officer to the Secretary of War, 1881* (Washington, D.C.: Government Printing Office, 1881), p. 165.
12. *Colorado Springs Gazette*, October 16, 1880.

CHAPTER TEN

1. Fetler, *Pikes Peak People*, p. 187.
2. Ibid., p. 191; ibid.
3. *Army Times, History of Signal Corps*, p. 64.
4. Gustavus Adolphus Weber, *The Weather Bureau: Its History, Activities, and Organization* (New York: D. Appleton and Company, 1922), p. 6.

CHAPTER ELEVEN

1. "Meteorological Observations," p. 470.
2. Ibid.
3. Ibid., p. 471.
4. Fetler, *Pikes Peak People*, p. 195.
5. Davidson, "Pikes Peak Prevaricator," p. 224.
6. Fetler, *Pikes Peak People*, p. 196.
7. "Meteorological Observations," p. 471.
8, Ibid.
9. Ibid.
10. Ibid.
11. Rupp, *Pikes Peak Duty*, p. 80.

12. U.S., Congress, Senate, Joint Commission to Consider the Present Organization of the Signal Service, Geological Survey, Coast and Geodetic Survey, and the Hydrographic Office of the Navy Department, with a View to Secure Greater Efficiency and Economy of Administration of the Public Service . . . , *Testimony,* 49th Cong., 1st sess., Sen. Misc. Doc. no. 82, 1886, pp. 85, 89.

13. *Army Times, History of Signal Corps,* p. 73.

CHAPTER TWELVE

1. "Meteorological Observations," p. 475.
2. Ibid.
3. As quoted in Rupp, *Pikes Peak Duty,* p. 139.
4. Ibid., pp. 143-44.
5. *Colorado Springs Gazette Telegraph,* October 18, 1964.
6. Ralph C. Taylor, "Colorful Colorado: Pikes Peak—Emblem of Eternity and Inspiration," *Pueblo Star-Journal and Sunday Chieftain,* December 9, 1962.
7. Fetler, *Pikes Peak People,* p. 229.
8. *Denver Post,* June 21, 1964.

BIBLIOGRAPHY

Abbe, Cleveland. "How the Weather Bureau Was Started." *Scientific American* 114 (May 20, 1916): 529.

Akers, Byron. "Believe It or Not: Someone Once Homesteaded atop Old Pikes Peak." *Colorado Springs Gazette Telegraph,* October 18, 1964.

Army Times, eds. *A History of the United States Signal Corps.* New York: G. P. Putnam's Sons.

Athearn, Robert G. *The Coloradans.* Albuquerque: University of New Mexico Press, 1976.

————. *High Country Empire: The High Plains and Rockies.* Lincoln: University of Nebraska Press, 1960.

Bancroft, Hubert Howe, *Works.* Vol. 25, *History of Nevada, Colorado, and Wyoming, 1540–1888.* San Francisco: History Company, 1890.

Bird, Isabella L. *A Lady's Life in the Rocky Mountains.* Introduction by Daniel J. Boorstin. Norman: University of Oklahoma Press, 1960.

Bliss, George S. "The Weather Business: A History of Weather Records, and the Work of the United States Weather Bureau." *Scientific American* 84, supplement no. 2172 (August 18, 1917): 110–11.

Colorado Prospector: Colorado History from Early Newspapers 22, no. 3 (July 1991).

Crouter, George. *The Majestic Fourteeners: Colorado's Highest.* Edited by Carl Skiff, photographed by George Crouter. Silverton, Colo.: Sundance Books, 1977.

Davidson, Levette Jay. "The Pikes Peak Prevaricator." *Colorado Magazine* 20 (November 1943): 216–25.

Eberhart, Perry, and Philip Schmuck. *The Fourteeners: Colorado's Great Mountains.* Denver: Swallow Press, 1978.

Fetler, John. Introduction by Marshall Sprague. *The Pikes Peak People: The Story of America's Most Popular Mountain.* Caldwell, Idaho: Caxton Press, 1966.

Great Britain. Meteorological Office. *Meteorological Glossary.* Compiled by D. H. McIntosh. London: H. M. Stationery Office, 1963.

Green, Stewart M. *Pikes Peak Country: The Complete Guide to Natural Wonders, Historic Sites, Attractions, and Outdoor Recreation.* Colorado Springs: Ponderosa Press, 1985.

Hansen, Harry, ed. "History of Colorado Springs." In *Colorado: A Guide to the Highest State,* 109–21. New York: Hastings House, 1941.

Harper's Weekly, November 8, 1873: 989.

Hawes, Joseph M. "The Signal Corps and Its Weather Service, 1870–1890." *Military Affairs* 30 (Summer 1966).

Holmes, Julia A. *A Bloomer Girl on Pike's Peak, 1858: Julia Archibald Holmes, First White Woman to Climb Pike's Peak.* Edited by Agnes Wright Spring. Denver: Western History Dept., Denver Public Library, 1949.

Hughes, Patrick. *A Century of Weather Service: A History of the Birth and Growth of the National Weather Service, 1870–1970.* New York: Gordon and Breach, 1970.

James, Edwin. *Account of an Expedition from Pittsburg to the Rocky Mountains Performed in the Years 1819, 1820,* 2. London: Longman, Hurst, Rees, Orme, and Brown, 1823.

Jessen, Kenneth. "The Prevaricator of Pikes Peak." In *Eccentric Colorado: A Legacy of the Bizarre and Unusual.* Boulder: Pruett Publishing Company, 1985.

Lapham, Julia A. "Storm Signal Service." *Earth and Air* 1 (February 1901): 3–5.

Lavender, David. *The Rockies.* New York: Harper and Row, 1968.

Marshall, Max L., ed. *The Story of the U.S. Army Signal Corps.* New York: F. Watts, 1965.

Mazzulla, Fred M. and Jo Mazzulla. *The First 100 Years: Cripple Creek and the Pikes Peak Region.* Denver: A. B. Hirschfield Press, 1956.

"Meteorological Observations Made on the Summit of Pike's Peak, Colorado, January 1874 to June 1888." *Annals of the Astronomical Observatory of Harvard College* 22 (1889): vii–vix, 459–75. Cambridge, Mass.: John Wilson and Son University Press.

Miller, Darlis A. "Pikes Peak Calling!: Sergeant Seyboth's Wintry Ordeal." *Colorado Heritage* (Spring 1991): 30–47.

Miller, Eric R. "New Light on the Beginnings of the Weather Bureau from the Papers of Increase A. Lapham." *Monthly Weather Review* 59 (1931): 65–70.

Moore, Willis L. "The Weather Bureau." *Scientific American* 53, supplement no. 1360 (January 25, 1902): 21802–3.

O'Keeffe, John Timothy. "Balaam of Pikes Peak." *Colorado Springs Gazette,* December 18, 1880.

———. "Pikes Peak, a Volcano." *Colorado Springs Gazette,* November 20, 1880.

_____. "Weather Bound." *Colorado Springs Gazette,* October 16, 1880.

O'Keeffe, John Timothy, and Eliphalet Price. "Attacked by Rats, Terrible Conflict on the Summit of Pikes Peak." *Pueblo Chieftain,* May 25, 1876.

Pike, Zebulon Montgomery. *An Account of Expeditions to the Sources of the Mississippi, and through the Western Parts of Louisiana, to the Sources of the Arkansaw, Kans [sic], La Platte, and Pierre Juan, Rivers.* Philadelphia: C. and A. Conrad and Company, 1810.

Rupp, Robert O. *Pikes Peak Duty: A Brief History of the Pikes Peak Military Reservation.* Fort Collins, Colo. R.O. Rupp, 1987.

Ruxton, George Frederick Augustus. *Adventures in Mexico and the Rocky Mountains, 1847.* New York: Harper and Brothers, 1860.

Shapley, Kathryn E. "The Weather Station on Pikes Peak." Unpublished paper.

Smith, James H. "Signal Station on Pikes Peak." In *Our Country: West.* Boston: Perry Mason Company, 1912.

Smith, Phyllis. *History of Floods and Flood Control in Boulder, Colorado.* Boulder: City, September 1987.

Sprague, Marshall. *Money Mountain: The Story of Cripple Creek Gold.* Boston: Little, Brown and Company, 1953.

_____. *Newport in the Rockies: The Life and Good Times of Colorado Springs.* Denver: Sage Books, 1961.

_____. *One Hundred Plus: A Centennial Story of Colorado Springs.* Colorado Springs: Colorado Springs Centennial, 1971.

Taylor, Ralph. "Pikes Peak: Emblem of Eternity and Inspiration." *Pueblo Star-Journal and Sunday Chieftain,* December 9, 1962.

_____. "Pikes Peak: Inspiration for Tall Tales." *Pueblo Star-Journal and Sunday Chieftain,* March 9, 1969.

Thompson, Philip Duncan, and Robert O'Brien. *Weather.* New York: Time, 1965.

U.S. Congress. Senate. Joint Commission to Consider the Present Organization of the Signal Service, Geological Survey, Coast and Geodetic Survey, and the Hydrographic Office of the Navy Department. . . . *Testimony.* 49th Cong. 1st sess. Sen. Misc. Doc. no. 82, 1886.

U.S. Department of Commerce. Records of the National Weather Service. Daily Journals. Observer Correspondence Series. Vol. 7 (1993–84), vol. 8 (1885), vol. 9 (1886–88), vol. 10 (ending December 31, 1896). Record Group 27. National Archives.

U.S. Department of Commerce. Weather Bureau. *Weather Glossary.* Compiled by Alfred H. Thiessen. Washington, D.C.: Government Printing Office, 1946.

U.S. Signal Corps. *Annual Report of the Chief Signal Officer of the Army to the Secretary of War.* 1873, pp. 281–82; 1874, pp. 57–58; 1874, Paper no. 6, "Summit of Pike's Peak, Colorado Territory" (October 1873–September 1874), pp. 113–26); 1875, pp. 59–60; 1876, pp. 61–62; 1877, pp. 64–65; 1878, pp. 68–69; 1879, p. 82; 1880, p. 68; 1881, p. 165; 1882, "Extracts Relative to Electricity, from Pike's Peak Monthly Abstract Journals," Appendix 72: pp. 893–96. Washington, D.C.: Government Printing Office.

————. *Monthly Weather Review.* 1883–present. Washington, D.C.: Government Printing Office.

U.S. War Department. "Practical Use of Meteorological Representations and Weather Maps." Washington, D.C.: Government Printing Office, 1871.

Vickers, W. B. "History of Colorado." In *History of Clear Creek and Boulder Valleys, Colorado.* Chicago: O. L. Baskin and Company, 1880.

Villard, Henry. *The Past and Present of the Pikes Peak Gold Regions.* Princeton, N.J.: Princeton University Press, 1932.

Weber, Gustavus Adolphus. *The Weather Bureau: Its History, Activities and Organization.* Service Monographs of the United States, no. 9. New York: D. Appleton and Company, 1922.

Whitnah, Donald Robert. *A History of the United States Weather Bureau.* Urbana: University of Illinois Press, 1961.

Wood, Stanley. "O'Keefe's Farewell." *Denver News,* December 12, 1881.

INDEX

A Note about the Author

Phyllis Smith is a historical writer living in Bozeman, Montana. She is currently investigating the history of the Gallatin Valley in that state.